Connecting Our World

GIS Web Services

Winnie Tang and Jan Selwood

ESRI PRESS

REDLANDS, CALIFORNIA

First printing June 2003. Second printing November 2003.

Printed in the United States of America.

Library of Congress Cataloging-in-Publication Data
Selwood, Jan.
Connecting our world : GIS Web services / Jan Selwood and Winnie Tang.
    p. cm.
   ISBN 1-58948-075-9
   1. Geographic information systems. 2. Geography—Computer network resources. 3. World Wide Web.
I. Tang, Winnie. II. Title.
G70.212.S46 2003
910'.285'4678—dc21   2003013719

Published by ESRI, 380 New York Street, Redlands, California 92373-8100.

Books from ESRI Press are available to resellers worldwide through Independent Publishers Group (IPG). For information on volume discounts, or to place an order, call IPG at 1-800-888-4741 in the United States, or at 312-337-0747 outside the United States.

Acknowledgments

GIS Web Services: Connecting Our World could not have been written without the cooperation and assistance of many people working in this developing field. We are grateful for their willingness to share their experience, enthusiasm, and expertise; for their comments and suggestions that have guided and enhanced the final product; and for the time they devoted to reviewing the material for accuracy.

The organizations involved in the case studies discussed in this book are named in each chapter. We acknowledge the tremendous assistance we have received from the individuals who provided advice and comments on behalf of those organizations, including: Kent Anness, WRIS GIS Manager, Kentucky Infrastructure Authority; John Callahan and his colleagues at Research and Data Management Services, University of Delaware; Tim Foresman for UNEP.net; John Spittal, Paul Lundberg and Dave Mole from Land Information New Zealand; Bruce Harold, Pasi Hyvonen and Harvey Wong from Eagle Technology Group; Robert Ford, GIS Manager at Crown Estate; Mark Jennings and Richard Bridgland from Chichester District Council; John Geary, Development Consultant, Wiltshire and Swindon Pathfinder; Rick Whitworth, Spatial Systems Administrator, City of Greater Geelong; Mark Taylor, Information Solutions and Innovations Manager, and Phil Beach, Government Information Access Services Manager at Department of Land Administration Western Australia; Gus Dominguez, Customware Manager, ESRI Australia Pty. Ltd.;

Mark Riddick, Chief Executive Officer, and Simon Coulthurst, Sales and Marketing Manager, NLIS Searchflow; Neil Johnstone, GIS Coordinator and Kelvin Hinton, Head of Development Services at Sevenoaks District Council; Roy Laming, Managing Director, and Peter Beaumont, Internet Services Manager, of ESRI (UK) Ltd.; Philip Nourse, Elvis Chan, and Alex Ng from Chesterton Petty Ltd.; Paul Yu from ESRI China (Hong Kong) Limited; Kolbeinn Gunnarson, Director of Development and Professional Services, Trackwell Software; Petter Nyborg, Geodata AS; Paul Overberg and Lynne Perri at *USA Today*; Daniel Day, General Executive, Membership Department, and Scott Johnson, Director of Graphics, Associated Press; Zhiquan Huang, North China Institute of Water Conservation and Hydroelectric Power; Gerry Ma, TIG Centre Ltd.; Tor Nielsen and Paul Forbess at IHS Energy Ltd.; May Yu and Erin Campbell from Homestore, Inc.

Staff at ESRI have been extremely supportive throughout, and helped in identifying issues and topics to be addressed, and in reviewing parts of the text. In particular we thank Joel Campbell, Scott Campbell, Carmelle Côté, Kris Goodfellow, Deane Kensok, Ian Koeppel, Ernie Ott, Paige Spee, Jonathan Spinney, Bernie Szukalski, Mike Tait, and Andrew Zolnai.

In the production of the book at ESRI China (Hong Kong) Limited, Diana Ip helped with general administration and Faheem Khan with the search and coordination of graphic material. At ESRI Press, R. W. Greene edited numerous versions of the manuscript, enhancing it in many ways, and, along with Christian Harder, guided it to publication. Doug Huibregtse designed the many iterations of the cover; Jennifer Galloway designed the book and did heroic graphics work. Edith M. Punt provided cartographic assessment. Tiffany Wilkerson did her customary meticulous copyediting and proofreading.

Data used for the creation of the globe images on the cover are courtesy of WorldSat International Inc.

REALTOR® is a registered collective membership mark which identifies a real estate professional who is a member of the National Association of REALTORS and subscribes to its code of ethics.

Preface

GIS WEB SERVICES combine two powerful enabling technologies: GIS, analyzing and integrating, and the Web, providing worldwide connectivity. The results of this synergy are widespread: data is easier to find, analytical tools can be shared in new ways, and both can reach a much, much larger group of people. What we are witnessing is a revolution in the way knowledge is shared, and created; GIS Web services are beginning to create a nervous system for the earth.

A number of related technological developments are both fueling, and being fueled by, these innovations. Spatial data servers have evolved so that data can now be modeled and stored in new and more flexible ways, and be served efficiently to very large numbers of concurrent users; since Web service architecture depends on robust servers, we can expect GIS will continue to become more server-centric. The Web has also brought a sharper focus to the power and efficiencies of standardization. Standards such as XML, SOAP, and UDDI offer the key to interoperability. They permit GIS packages not only to exchange data and services with each other, but also connect and exchange services with virtually any computer application.

The potential for processes such as these is unlimited. A user commencing a project can come to the Internet and identify remote services that provide both the necessary data and the modeling functionality. These can be integrated within the user's own systems. The user's system

can trigger data hosted on one or more remote servers to be passed to models on another remote server, with the result being returned in a form that can be directly integrated with the original user's existing applications. There is no need to store or develop the data or services accessed locally; they can be consumed as and when required, and are invisible to the end-user client.

GIS Web Services: Connecting Our World explores this technology, illustrating it with a range of examples drawn from organizations around the world, who are both connecting that world, and blazing a new trail for GIS.

▶ *Jack Dangermond*
President, ESRI

THIS BOOK EXPLORES a technology of profound importance to the way information management systems now and in the future are designed. GIS Web services offer the ability to build robust applications and to share and integrate data and functionality over the World Wide Web—applications that are virtually platform-independent, easy and cost-effective to maintain and update, and which can be accessed by a potentially global audience. GIS Web services present an opportunity to dismantle some of the artificial barriers that limit integration and communication among information systems, whether these have been created by the restricted, vertical nature of data storage or application design, or by the incompatibility of data formats, hardware, or software. GIS Web services also break down the barriers of distance, permitting users anywhere equal access to a wealth of data and services.

For my country, the People's Republic of China, this is especially significant. With the world's largest population, almost 1.3 billion people, a rapidly developing economy and a natural environment of great diversity, richness, and beauty, we have an enormous responsibility to our descendents to manage wisely. This has motivated the development of essential GIS applications and long-term, productivity-enhancing GIS application programs. These have as their goal rationally utilizing resources, protecting the environment, developing the economy, and raising the living standards of the general population. Substantial human and financial resources have been invested to develop the National Spatial Data Infrastructure (NSDI), and more and more geospatial data is being provided to, and being shared by, a variety of users.

We are constantly reminded of the importance of interconnections and relationships in our world and how little we understand these. GIS Web services, including many of the examples in this book, help to share data and technology across barriers of technology and distance. This is good: it promotes learning, understanding, sharing; it demonstrates how GIS and geospatial technology can make a positive difference that affects all of us, wherever we live.

▷ *Chen Jun, Professor in GIS*
   President of China Association of GIS
   President of ISPRS Technical Commission II
   President of National Geomatics Center of China

Contents

GIS Web Services
Connecting our world

FUNDAMENTALLY, GEOGRAPHIC INFORMATION INTEGRATES. Everything, every feature, process, and phenomenon has a location; often, location is the only obvious connection among a huge range of disparate observations and information that we receive every day. Spatial relationships bring them together and help us order them.

Because of its universal nature and its ability to integrate, geographic information and reasoning are deeply embedded in the way we understand the world around us, the events within it, and how we make decisions based on this understanding. Whether the decision concerns finding the fastest route home through rush-hour traffic, or when and where to plant a particular crop, or what kind of dam is required on a particular river, or how best to manage the evacuation of a village threatened by devastating mudslides, location information is deeply integrated into the decision-making process. Frequently it is so well integrated in the way people store, view, think, and analyze information that they are unaware of its underlying geographic nature.

One effect of the ubiquitous nature of geographic information is that although different users may apply it for countless different purposes, the information itself and the way in which it is assembled and analyzed are often common, and shared. For example, schools, government departments, emergency services, transport engineers, utility companies, planners, lawyers, businesses, and commuters within a particular area are

all likely to be using very similar, even identical, coastline, topographic, administrative, boundary, address, transportation network, or point-of-interest databases. They use this information for widely differing purposes, but the basic information is the same. Furthermore, the functionality they use in amassing and analyzing it is shared as well—they all need to display maps, find addresses, measure distances and areas, overlay layers of different information, and undertake proximity searches.

Representing, storing, and analyzing geographic information has, at least until relatively recently, been a struggle. Obviously, geographic detail has been recorded on maps and atlases for thousands of years. But hard-copy maps and atlases have tended to separate the geographic from other forms of data (textual or tabular), and to make the capture, analysis, and maintenance of spatial data a specialized, exclusive skill. The development of GIS is a response to this, and GIS has been driven by the need to find ways of representing and working with spatial data that allow that data to play its natural integrating role—ways that break down the unnatural separation between spatial and nonspatial data, that allow spatial analysis to be seamlessly integrated with other technologies, to allow both analyses and data to be shared. GIS Web services represent an important latest step towards achieving these aims.

GIS WEB SERVICES

GIS Web services have the potential to revolutionize the way in which GIS is developed, accessed, and used. They make it easier to share geographic data and functionality, and for GIS to be deeply integrated into other technologies.

A Web service is a software component that can be accessed through the World Wide Web and used by other applications. GIS Web services provide spatial data or functionality on the World Wide Web. They make it possible for users to access GIS data and functionality through the Web and to integrate them with their own systems and applications without the need to develop or host specific GIS tools and data sets themselves.

This is a significant development, and may turn out to be a fundamental shift in the way GIS develops. GIS Web services make it feasible for multiple organizations that need to access the same data to do so from a single database hosted as a Web service, rather than (as is often the case now) simply duplicating the data in each organization. They make it feasible for a spatial function, such as map generation based on user-specified addresses, to be shared by multiple organizations, and to be plugged straight into an application with little or no development. Organizations using the function do not even need to store spatial data or software on site. The same *generate map* function can be used by many different organizations across the world and in many different types of software.

Using GIS Web services, organizations developing and maintaining spatial data or functionality can share their products and services with a worldwide, Web-connected community. Those using or consuming spatial data or services can do so without incurring overhead locally.

PRACTICALITIES

The use of GIS Web services will have profound savings in both time and labor for virtually every organization, even in a simple application. An obvious example is that of a typical municipal government where various different departments scattered across town all need to access the same basic data sets—such as roads, buildings, addresses, planning zones—and need the same functionality—navigating the map, generic spatial queries, printing hard-copy maps. Of course, each department may have its own specific data sets or functions that only it uses, but if each department can integrate these specific demands with a Web service that hosts shared data and functionality, it not only reduces development and maintenance overheads, it helps ensure consistency across the whole organization.

Consider the example of a slightly more complex application, a dispatch service operated for small- to medium-sized delivery companies. Each company manages delivery orders in different software packages. Normally when delivery trucks are ready to roll, these systems generate a list of addresses that are given to the driver, who calculates the most effective route between stops.

Introducing a GIS Web service could streamline this process greatly. The address lists would be passed to a remote dispatch service that could generate the most efficient delivery route. The dispatch service would receive a list of addresses, geocode these against its own up-to-date street and address databases and then run a routing algorithm to generate the most efficient route. Taking this one step further, before returning the recommended route to the delivery company, the dispatch server could automatically pass the addresses through a second Web service run by a local traffic center, which would check for any known road closures or traffic problems on the proposed route. Based on the response from the traffic center Web service, the dispatch service could rerun the routing model and search for alternative routes, or recalculate route timing. The result, a systematically defined dispatch route that takes into account current and forecast traffic conditions, would then be sent as route maps or driving instructions back to the delivery company, or even to a laptop or PDA in the driver's cab. There would be no need for the delivery company to manage and maintain its own spatial data sets and routing models, or to rely too heavily on the street-wise knowledge of their drivers. The service provides a single one-stop shop—up-to-date address and road networks, up-to-the minute traffic condition reports, and state-of-the-art routing.

The actual software running the dispatch and traffic services can be written in entirely different software languages working on different operating systems, but they can work together as a single entity. Many delivery companies can access the service at the same time, even though all may be using different software to manage their orders and addresses.

What makes GIS Web services work is their conformity to open, widely accepted interoperability standards that allow data and, more importantly, commands, to be passed among systems linked by no more than an Internet connection. By focusing on standardizing the way that data and commands are exchanged among computer systems, GIS Web services introduce a whole new level of interoperability.

Systems that host and use Web services do not need to know anything about each other—what software is providing the service, what operating system it sits on, nor what application will use it. All that is needed is a connection to the Web and support of Web service standards. This is a huge advance over earlier attempts at integrating applications; many of these required that network connections be predefined and static, and that both the service provider and client be built from the same software. Web services use the Web to provide straightforward, universal access anywhere, at any time, providing truly distributed computing. In fact, a better term might be "collaborative computing."

Not surprisingly the possibilities of Web services are generating considerable interest in the IT industry—some market research estimates the global market for Web services will rise from $1.6 billion (U.S.) in 2004 to $34 billion by 2007.

GIS Web services offer a new framework through which geographic data and functionality can flow. Because it is so flexible, the Web services architecture can be scaled to support the delivery of GIS services for any size—for small companies, businesses and schools, across governments and multinational organizations, to the provision of data and functionality to a truly global audience.

THE WEB SERVICES PROCESS

Web services are based on the technology and standards that have evolved over the last twenty years as the Internet and the Web have developed. The underlying process is fairly basic: when a provider is ready to release a new service, it publishes details about what the service does and how to interact with it. Service consumers—which can be clients, applications, or other services—identify appropriate services across the Internet, either by going to known URLs or by searching service catalogs or registries that act like the Yellow Pages, and which provide details of published services. Once identified, the consumer uses the service, sending a request and receiving the response again over the Internet.

The key point is that the Web service process concentrates solely on how data and commands are passed from one computer to another. The applications at either end of the Web-service process—the ones that actually do the work of providing a service, or using it, are essentially irrelevant. Web services provide a framework for describing how to pass commands to a particular application and how to understand its response. This means that virtually any application can be published as a Web service—whether it is an application developed specifically as a Web service, or part of a legacy system that has been around for ages.

The framework that permits this to happen depends entirely on open, published standards.

THE STANDARDS BEHIND WEB SERVICES

Web services are made possible by the widespread acceptance of standards, perhaps the most important of which is eXtensible Markup Language (XML). All the Web services described in this book use some form of XML.

To understand the benefits of XML, it is necessary to understand HTML, HyperText Markup Language. Developed in 1991, HTML revolutionized Web page design. It provided a standard way to describe how data in a file should be displayed. Unknown, remote Web clients could receive an HTML-formatted document across the Internet, read it, and display it exactly as it was intended by the document's author, regardless of differences in software or operating system. HTML made it possible to design exciting, user-friendly Web pages that could be easily viewed and navigated.

HTML remains the primary way of formatting Web pages. But by the mid-1990s, its limitations were becoming apparent. Although providing a very clear standard, HTML is rigid. Transmitting data that does not fit the generic document format for which HTML was designed is difficult and inefficient—data such as GIS layers with associated attributes, symbology, or complex graphics. And, aside from hyperlinks, HTML cannot be used for meaningful interaction between the user and remote Web server, making it very difficult to support an application any more complex than data presentation.

XML was proposed as the solution. XML was designed to complement rather than replace HTML, and differs from HTML in two important ways. First, XML describes the data contained in a document; HTML describes only how to present data contained in a document. An XML-formatted document can therefore carry much more than pieces of data to be displayed on-screen—it can also contain commands and parameters to be passed to a remote application, or complex data sets returned from an earlier query.

Secondly, XML is extensible—meaning that unlike HTML, which was rigidly defined to describe a single generic type of document, XML permits users to write their own definitions, so that virtually any type of data can be described. As long as these document definitions, called schema, conform to XML schema standards, they can be transferred with the document and will ensure that client applications can understand the data contained within it. This means that industries or applications that have particularly complex data sets can write their own XML-compliant data definitions.

GEOGRAPHIC MARKUP LANGUAGE

▷ The Open GIS Consortium (OGC) has led the development of Geographic Markup Language (GML), an XML schema designed to provide a cross-platform description for spatial data. Since its launch in March 2001, this effort has been gaining wide support within the GIS community. The standard is still evolving, and work within the OGC by leading GIS and IT vendors such as ESRI, Oracle®, Sun Microsystems™, and key users is continuing, with the goal of enhancing this standard to allow for complex commands and very large data sets. OGC is on the Web at *www.opengis.org*.

XML immediately added great flexibility to Web-based data transfer— XML-formatted data can of course be used for presentation, but it can also be used to transfer commands and parameters to trigger remote systems to run applications, and to transfer complex data sets—vector data and related attributes, or parts of databases. In 1998, the first XML specification was agreed to by the main standards organization of the Web, the World Wide Web Consortium (W3C).

XML simply describes the data contained in a document—it does not address how this document can be transferred from one system to the next. For data to be transferred between computers there must be some agreement as to how communication will take place—such agreements are called protocols. In practice such transfer protocols are simply statements at the beginning and end of an XML statement that provide information about what the document is and how it should be used. There are many protocols in use; some are specifically designed to transfer messages between systems using similar software products; others are agreed international standards designed to be recognized and used by any software platform. ArcXML is one example of the former, an open published standard adopted by ESRI to transfer data between ESRI applications and other GIS systems that can translate or interpret the ArcXML protocol.

Simple Object Access Protocol (SOAP), written in XML, is perhaps the most commonly used protocol standard, and is designed to be entirely application-independent. Web services that conform to the SOAP standard can be integrated with any other Web-based application.

The World Wide Web Consortium is found on the Web at *www.w3.org*.

ESRI WEB SERVICES DEVELOPMENT

▸ ESRI has been cognizant of the significance and potential of serving geographic data and functionality across the Web, and has been a pioneer in the development of GIS Web technology and services. In late 1996, ESRI first delivered Internet mapping extensions for MapObjects®, a component-based solution for developers, allowing users to extend GIS capabilities to the Internet; other extensions allowed the same capability for ESRI's desktop GIS product, ArcView®. Also released was ArcExplorer™, a geographic data viewer that could access Web-mapping services, and which later evolved into both a browser and Java™ version; all versions were available free of charge.

At about the same time, ESRI also began work on the development of the REALTOR.com® and Visa.com sites, each providing a challenge to developers to meet the needs of a high-volume site with specific architecture needs. The technology used for these was released in 1997 as ArcIMS®, now ESRI's flagship platform for GIS Web mapping.

With ArcIMS, XML was adopted as a core standard protocol, even before it was ratified by the World Wide Web Consortium; this open standard was adopted throughout ESRI's Web technology. The company was also an early adopter of the SOAP standard. The year 1998 saw the release of ArcData℠ Online, the mapping and data Web service that permitted users to browse and create maps across the Internet. The collaborative, multi-participant Geography Network℠, built using ArcIMS, and featured in chapter 1, was launched in 2000. The ArcXML schema was published the same year. In 2002 came ArcWeb Services, which provided hosted services—data and functionality—that could be incorporated into any application or Web site. Other hosted services that came online that year included MapShop (discussed in chapter 9), and the robust service packages available from ESRI Business Information Systems, ESRI BIS. Most recently, ArcWeb Services for Developers provides a component-based toolkit of hosted data and services for application development, and ArcWeb Services for ArcGIS Users, providing access to individually hosted data or services.

UDDI can be found on the Web at *www.uddi.org*.

Two other standards also deserve mentioning at this stage. The W3C has adopted Web Service Definition Language (WSDL) as a standard way of describing Web Services. WSDL, like SOAP, is based on XML and is used by service providers to publish details of the services they are offering. Going hand-in-hand with WSDL is Universal Description, Discovery and Integration (UDDI). UDDI is promoted by a group of leading computer companies and provides a standard way for defining Web service registries so that services can be easily searched and selected. Though neither WSDL nor UDDI have gained as much acceptance as XML and SOAP have, they are increasingly being adopted as the standards for published Web services and establishing Web service directories.

Putting these standards into the Web services process would mean that for a service designed to be available to any Web application, its provider would develop a description of it using WSDL and publish this to a UDDI directory. The client application would find the service through the UDDI directory and use the WSDL description to establish how to communicate with it. Requests and responses between the service and the client would be passed in SOAP-wrapped XML documents.

Where a service is designed to be used by a specific community or group that knows where it is located and how to communicate with it, there is of course no need to publish details in WSDL or on a UDDI directory. Client applications simply go directly to the service URL and communicate either using SOAP- wrapped XML or some other protocol common to both client and server, such as ArcXML.

INNOVATORS

GIS Web services can range from single functions such as distance calculations and simple spatial search routines to complex map generation, spatial editing, and modeling. Some are made available freely, others are run internally on corporate intranets or are established as commercial services. This book provides a selection of how different organizations are beginning to explore the potential of GIS Web services. The range and scope of these examples is broad, but in no way comprehensive.

Many examples drawn from local and national governments, in New Zealand, Australia, the United Kingdom, and the United States, demonstrate how Web services can facilitate public access to very large databases held by public organizations. Others, such as the system established by Chesterton Petty Ltd. in Hong Kong and those based on Eview in the City of Geelong, Australia, and on MapsDirect in the United Kingdom, show how Web services—either running on a local intranet or accessing externally hosted databases—can improve management and access to data sets within organizations both large and small. The chapter on IHS Energy shows how commercial data providers are beginning to make use of Web services to facilitate access to their databases and to improve customer support. The National Land Information System Searchflow system in the United Kingdom, the Homestore® real estate locator, and the MapShop map generation tool are all examples of Web services that provide

complete packaged solutions for particular markets. Examples from Iceland and Norway show Web services being integrated with other technologies to provide mobile tracking and location services to the telecommunications industry. An example from a university in China shows the potential of using Web services to help to teach and disseminate knowledge of GIS itself.

GLOBAL WEB SERVICES

▸ Jack Dangermond, founder and president of ESRI, talks of GIS Web services extending to form an information network that will become a kind of nervous system of the world. The metaphor is not hard to follow—GIS can model and store geographic data, integrate, analyze, and reveal meaning through location, and is coupled with Web services, which enable connections among users, systems and databases, in real-time, concurrently, across the world. The Web provides the connectivity that enables global access to geographic databases and functionality where and when it is required, thus realizing the concept of a truly societal information system.

The Geography Network
A framework for sharing spatial data

1

THE GEOGRAPHY NETWORK was launched at ESRI's International User Conference in July, 2000. At a time when the focus of many in the IT and GIS industries was on centralizing data in ever larger data warehouses, a Web portal service that permitted spatial data to remain in one location and yet be shared among organizations was seen as somewhat radical. However, the Geography Network has proven its worth many times over, and its basic structure has been adopted by many organizations, large and small, as means of sharing and disseminating spatial data and functionality. The direction charted by the Geography Network is entirely consistent with that of Web services, as well as with the trend toward loosely bound, cooperative computing.

The Geography Network was established as a metadata search-and-discovery framework that permits exploration of distributed spatial data sets and services. Highly flexible and scalable, it has been adopted as the model for data sharing in local and national government initiatives, as well as by corporate and international organizations, and now provides access to a vast range of commercial and public-access data sets and services. It has evolved to provide not only the means for data discovery and visualization, but also the capability to serve data and targeted services directly to client applications.

DATA: HOW TO CAPTURE IT, FIND IT, DISSEMINATE IT

Finding, capturing, and disseminating data have been constant concerns for anyone involved in GIS over the last twenty years. In the early days, data was frequently just not available in digital form—if it was needed it first had to be digitized. As more and more GIS databases were compiled, the principal concern shifted to how best to disseminate and share data. The nature of geographic information means that although individual departments or organizations may compile and use some unique data sets relevant only to themselves, more often than not they share a common interest with other organizations in common geographic information, such as topographic base mapping, street plans, census statistics and the like.

All too often however, the complexities of data transfer, update, and format have meant that data has remained locked in the databases of individual departments or organizations—frequently with only a relatively small group of regular users fully aware of the data held. At the same time, the importance of shared, up-to-date spatial data became an ever-higher priority. The increasing integration of GIS into the decision-making processes of government, administrative, academic, and commercial organizations highlights the importance of ensuring that everybody is working from the same consistent data sets. Increasing adoption of GIS within disaster-response and homeland-security plans demonstrated the importance of being able to search quickly for, and integrate data sets in, new and unanticipated ways.

From the mid-1990s, initiatives to improve the way in which spatial data is disseminated and shared have been pursued energetically. The Federal Geographic Data Committee (FGDC), the International Organization for Standardization (ISO), and European Union have all attempted to bring order to the way data content, purpose, and origin is described through metadata standards. Leading GIS vendors either independently or by participation with major user organizations in the Open GIS Consortium (OGC) have also been successful in expanding the availability of data in open, fully documented formats. In addition, a number of national spatial data infrastructure projects have been implemented by governments around the world as a means of promoting access to their own data resources.

The Geography Network can be seen as building on these developments to provide a multiparticipant framework that facilitates discovery and use of the spatial data and functionality.

The concept behind the Geography Network is to establish a Web-based catalog or registry of Internet services, providing spatial data, maps, or functionality. Rather than users having to check each and every possible service provider, they can use the Geography Network as a one-stop shop through which they can search for the service or data set that meet their particular requirements. In some respects, it acts a bit like a Yellow Pages telephone directory, enabling spatially related data or functionality to be registered with a central catalog. However, unlike the Yellow Pages, services can be cataloged and searched based on their location, so users interested in a particular area will retrieve information on only those services that are relevant to that area.

The concept behind the Geography Network conforms to Web services concepts of Publish, Find, and Use. The Geography Network provides a metaportal through which data can be discovered and explored, and access to services that enable data can be downloaded, purchased, and used within Web applications such as ArcExplorer or as a direct service to ArcGIS® applications.

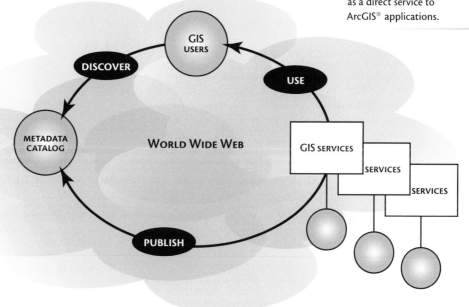

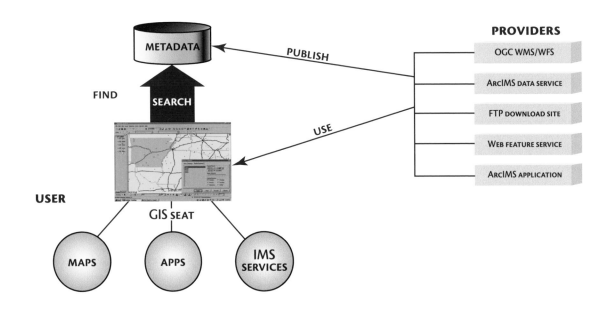

The Geography Network
is on the Web at
www.geographynetwork.com.

The Geography Network is freely available and open to any organization that wishes to participate, including data publishers or commercial organizations with services to sell, and those that provide services free of charge. The only requirements are that the service must provide spatial content—maps, spatial data sets, or spatial functionality—and come documented with a brief metadata description. Services include Web pages or FTP sites from which data, maps, or application tools can be downloaded as static information, and links to hosted Web services that provide direct access to dynamic data and Internet mapping services. ArcIMS data and application services and OGC's Web Mapping Server (WMS) services are supported, and there is no limitation on the format of the static data that can be registered.

FGDC can be found on the
Web at *www.fgdc.gov.*

The metadata description is based on the minimum requirements of the FGDC standard. This includes a broad summary of the contents, purpose, update frequency, access, and data type description. In addition, it includes the spatial domain—or the areal coverage of the service. This information provides the key to the cataloging service. With this, the user can search registered data and services, either with textual queries for the type of service, data, provider or place name, or through dynamic spatial searches that permit an area of interest to be defined, from which any service that falls within that location can be retrieved.

The beauty of the Geography Network is that it is highly scalable. It can be established to catalog information and services within particular organizations, cities, states, national, or international initiatives. By providing a standard architecture, these catalogs can be easily combined, so that a catalog created on the Geography Network and which provides a gateway to data relevant to a particular city can be integrated with that of a state or a nation. In this way, a reference network grows organically as more organizations participate and develop catalogs of their own data sets and services. As the network extends, so does the volume and richness of content that can be searched.

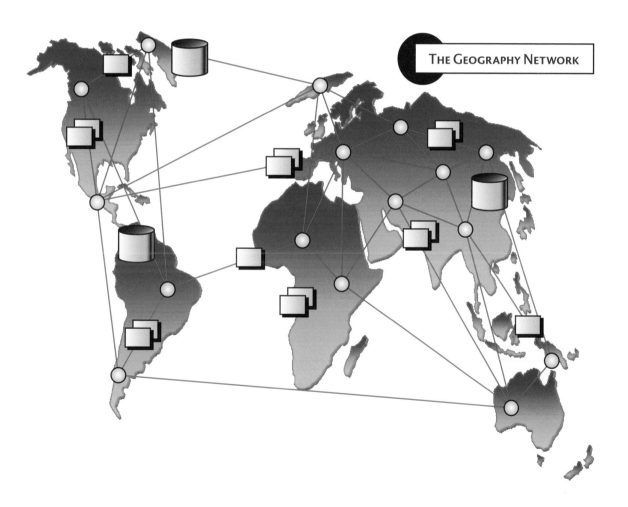

THE GEOGRAPHY NETWORK

The Geography Network, established in 2000, now spans the world and has become one of the first stops for people commencing a new plan, project, or journey, and who are looking for spatial data sets.

The Geography Network portal managed by ESRI provides a single point of access to many terabytes of data and to many services hosted by both commercial and noncommercial organizations. There are more than three hundred service providers registered, ranging from university research teams and local city governments to state portal sites and those of major national and international bodies such as NASA, the United States Geological Survey (USGS), and the United Nations. Many of the data sets and services provided by these organizations can be utilized, linked to, or downloaded free of charge; others are available for a fee or by subscription, paid directly to the service provider.

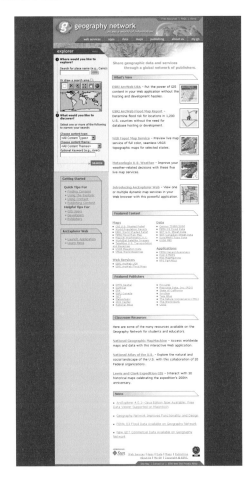

The range of information provided is vast. If you are interested in the border area between China and North Korea, the Geography Network will return more than 190 different maps, data sets, and application services covering the region. Many of these are global data sets from organizations such as the World Bank, United Nations, or the National Imagery and Mapping Agency (NIMA), and provide general topographic data as well as data sets that encompass more than physical geography, such as average life expectancy, levels of GDP, and ecoregions. Others, from specialized scientific organizations such as the Smithsonian Institute and NASA provide data from their unique databanks: volcanic activity, for example, or detailed coastal and freshwater temperature data. Other data sets offer reasonably large-scale mapping of a region, such as the National Bureau of Surveying and Mapping of the People's Republic of China's 1:100,000 topographic maps; a number of services offer high-resolution Digital Terrain Models (DTMs) and satellite imagery.

The Geography Network enables data to be searched and previewed and, in the many cases where data is hosted as a live service, loaded directly into the ArcExplorer Web application. This provides a sophisticated Web map interface that allows users to view multiple layers, set transparency levels, to query, and print.

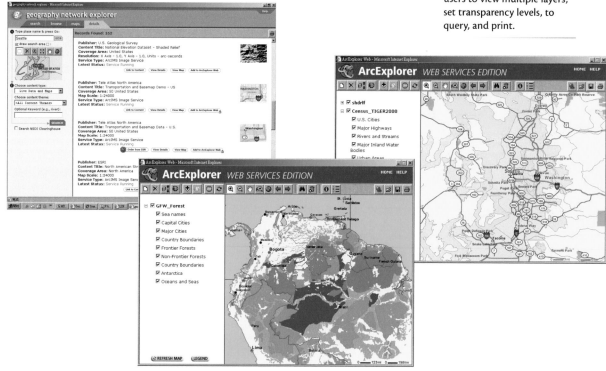

Other parts of the world are covered in even more detail by even more services. Undertake a search on say, San Diego, California, for example, and there will be more than four hundred references returned. These include links to San Diego city's and county's own geographic services, providing Internet mapping or data downloads of highly detailed maps at scales of 1:1,000 or better. The majority of the data sets can be viewed online or can be downloaded or integrated directly with users' GIS systems.

The hardware and software architecture of ESRI's portal not only acts as the primary catalog for Geography Network, but also serves more than 600 GB of data from ESRI and other providers, either as downloadable maps and data sets, or as dynamic data that can be integrated into ArcGIS, ESRI's desktop and workstation GIS technology. The system generates more than 1.4 million maps a day, with peak transactions at around 150,000 an hour. The same architecture is used to host ESRI's SOAP-based ArcWeb services, including the freely available Place Finder service that is part of the ArcWeb for developers service bundle. The latter combines access to state-of-the-art standards-compliant GIS Web services (providing tools for map and image visualization, routing, proximity searches, and the like), as well as cost-effective access to commercial data sets. Data is stored on two Sun Ultra Enterprise servers holding more than three terabytes of space for data, and there are currently four Sun Enterprise application servers and between eighteen and twenty-four machines acting as Web or Internet Map servers, depending on the load. The hardware is split between two locations to guarantee twenty-four-hour-a-day service.

A number of gateways have been implemented based on the Geography Network architecture. They provide access to spatial data resources for the states of Kentucky and Delaware, and for the United Nations Environmental Program.

COMMONWEALTH OF KENTUCKY'S KYGEONET

Kentucky's Geography Network (KYGEONET) provides a good example of the ease with which data services provided by a number of organizations can be integrated through the Geography Network architecture to serve the local community. There are many organizations actively developing GIS applications within the Commonwealth of Kentucky and many have adopted Internet mapping as a means of promoting access to their data sets. The Kentucky Office of Geographic Information (KYOGI), Kentucky Geological Survey (KGS), Kentucky Water Resources Information System (WRIS) and the Kentucky Natural Resource and Environmental Protection Cabinet (NREPC) all established sites that enabled data to be directly browsed or downloaded. Originally, none provided keyword or geographic search capabilities and, for the user, finding the location of a particular data set could be a challenge.

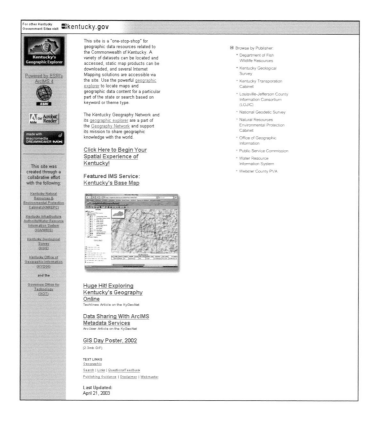

KYGEONET is a collaborative effort providing access to metadata, data download facilities, and online mapping services that provide access to a wealth of geospatial data across Kentucky.

With the support of the Kentucky Geographic Information Advisory Council, data providers got together to establish a local Geography Network—the KYGEONET—that could be used to search for data across multiple organizations. Within a week of its launch, some forty-four data services had been registered with the server; within six months, there were 140. The means of serving data, through Web sites, FTP downloads, or dynamic data transmission, remain the responsibility of the organizations providing it, and many have simply maintained existing systems and submitted standard metadata descriptions. KYGEONET simply makes these easier to find, search, and use. All data is provided free of charge as a service to the community. KYGEONET provides a simple, easy-to-use, one-stop search engine for the commonwealth's geographic data resources. Once established, the system was registered as a node of the Geography Network, making it a part of a worldwide spatial information network.

KYGEONET can be found on the Web at *kygeonet.state.ky.us.*

KYGEONET runs on one spatial database server running Microsoft® SQL Server™ and ESRI ArcSDE® 8.2, which hosts all metadata and common data layers. Two Internet servers are used. The primary server hosts metadata and gazetteer services and provides general image, base, and thematic services, while the secondary server is used for load balancing. Internet servers run ArcIMS 4.0, Microsoft Internet Information Server (IIS) 4.0 and New Atlanta's Java Servlet Engine. A T1 link provides access to the Commonwealth's Wide Area Network (WAN).

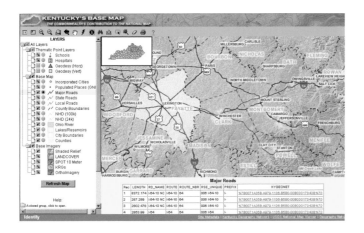

KYGEONET map viewing service hosts a variety of data sets and allows users to undertake searches and identify queries on the underlying tabular data. Links provide access to related data sets and photos. Scale sensitive orthophoto, vector, and map services turn on as the user navigates the map.

DELAWARE'S DATAMIL

The Delaware Data Mapping and Integration Laboratory (DataMIL) has established its Geography Network architecture as a means not only for sharing and facilitating searches of data, but also for promoting feedback and updates of data sets from users. The DataMIL acts as an interactive, online collaboration to facilitate collection, integration, maintenance, visualization, and distribution of geographic framework data for the State of Delaware. As in Kentucky, collaboration among a number of organizations made the system possible—in the case of Delaware among the University of Delaware, the Delaware Geological Survey, the United States Geological Survey (USGS), and the Delaware Geographic Data Committee (representing state agencies, and county and municipal governments). DataMIL serves the complete Delaware Spatial Data Framework, which includes topography, aerial photography, administrative boundaries, tax parcels, and road and river networks—all base topographic and administrative data layers used by municipal and state authorities.

The DataMIL provides a single point of access to explore Delaware's geospatial data resources. It provides a vehicle through which data can be shared among organizations, visualized, downloaded, and maintained. The DataMIL Portal provides access to the MapLab, Metadata Explorer, Data Integration Laboratory and Discussion Forum.

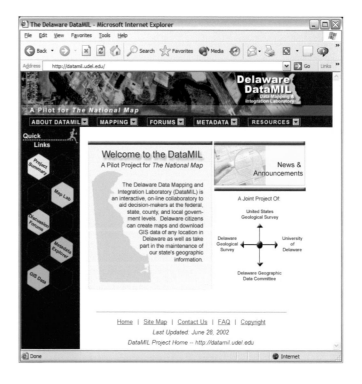

The DataMIL system extends the basic Geography Network architecture to provide a number of distinct Web-based applications, the primary ones being the Portal, the Map Production Laboratory (MapLab), and the Discussion Forums. All users enter the system through the Portal, which provides background on the Delaware Spatial Data Framework, development and maintenance notes, and access to information regarding all facets of the project. The MapLab is a sophisticated HTML-based interface for browsing and map creation. Metadata information is built directly into the MapLab, as is the ability to create printable (PDF or JPEG), real-time, topographic maps, and vector data downloads. The Discussion Forums provide a focus for discussion, both on the system itself and the data it serves. For example, if a user has a problem browsing the site or finds an out-of-date feature on a map, that user can post the question or comment to a forum for review and, if deemed appropriate, correction by the organization responsible.

Two supporting applications are the Delaware Metadata Explorer (ME) and the Data Integration Laboratory. The Delaware ME provides a Web-based search engine to the Delaware Spatial Data Clearinghouse, Delaware's node on the National Spatial Data Infrastructure (NSDI). The Data Integration Laboratory, based on the ArcIMS Java Viewer, provides online editing and mark-up facilities for data stewards.

The start page of the Delaware DataMIL system.

Delaware's distinctive MapLab interface provides a truly intuitive environment in which data services can be assembled and browsed. It provides all the basic map navigation tools, the ability to annotate maps, and a range of helpful features such as predefined scale and zoom settings, image capture, and map creation. In addition, data within the view area can be easily extracted and downloaded along with related metadata files.

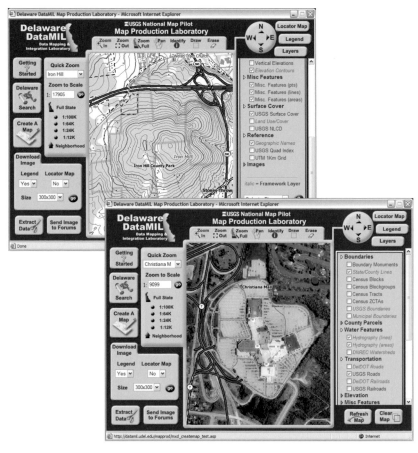

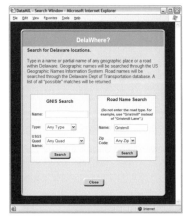

In addition to the Web-based interfaces, DataMIL hosts numerous ArcIMS image and feature Web map services for direct connection through a GIS. The map services include the Delaware Spatial Data Framework, historical aerial photography, and several Delaware census TIGER® and demographic layers. All image services are OGC WMS-compliant, while feature services are OGC WFS-compliant. Likewise, DataMIL's metadata Web services support both ArcIMS AXL and ISO's Z39.50 requests. This kind of collaborative system, open and publicly accessible, is the reason why DataMIL is one of the leading prototypes through which concepts for USGS's National Map and Federal Geospatial One-Stop initiatives are being explored.

Currently, DataMIL is being used extensively by state, county, and local agencies as well as by the University of Delaware's faculty and students. The DataMIL servers are located at the university and are producing nearly three thousand maps per day. To keep the data as current as possible, data stewards update DataMIL through an automated FTP procedure whenever modifications are needed. Five machines make up the DataMIL hardware configuration. The database server, on a Dell® PowerEdge 6400 with four 700-MHz processors and 4 GB RAM, utilizes Oracle 8*i*™, ArcSDE and ArcGIS Desktop. The Web and map servers are each on a Dell Precision 530 2.4-GHz dual-processor machine with 2 GB RAM, and run Sun ONE Web Server, Sun ONE ASP, and ArcIMS. Additional machines provide firewall protection and an FTP site.

DataMIL is on the Web at *datamil.udel.edu.*

NATIONAL MAP AND GEOSPATIAL ONE-STOP

▶ The National Map is a concept being promoted by the United States Geological Survey, and is an initiative designed to create a robust base map framework for the United States. The USGS maintains the primary series of topographic maps in the United States. Two of the key aims of The National Map are to develop creative partnerships that will ensure that this base map data is kept current, complete, and consistent throughout the country, and also to help make data accessible to a national audience whenever and wherever it is needed. Participation in local initiatives such as the Delaware DataMIL is seen as an essential step in achieving both goals. More information on The National Map can be found at *nationalmap.usgs.gov.*

Geospatial One-Stop is a major e-government initiative sponsored by the Federal Office of Management and Budget. The initiative recognizes the importance of consistent reliable geospatial information to effectively carry out the business of government. It aims to promote access to geospatial information throughout the United States, to facilitate sharing of information and collaborative partnerships both in standardization and data set development, and to promote the National Spatial Data Infrastructure. More information on the Geospatial One-Stop can be found at *www.geo-one-stop.gov.*

INTERNATIONAL AGENCIES

The Geography Network is also bringing together distributed data sets from a range of international aid and development organizations and think tanks, including the World Bank, United Nations Environment Program (UNEP), the World Wildlife Federation (WWF), and the World Resources Institute. One of the important roles of such organizations is the collection and publication of data on a global scale, often beyond the scope of any single government: trends in global population, in poverty and life expectancy, the speed and nature of change in environment and habitats and their implications, biodiversity, and the effect of global climate change. How to distribute the results of their work, to ensure easy, open access to users from around the world has been a constant challenge. Many already maintained comprehensive Web sites themselves, but few provided expressly spatial search or display capabilities. Linking to the Geography Network not only provides this, it facilitates the integration and comparison of data sets from other related organizations.

In September 2000, UNEP designed UNEP.Net as a means of bringing together the wealth of scientific information that existed both within individual programs sponsored by the U.N. as well as with many other organizations and bodies working on the global environment. A team of thirty U.N. staff drawn from different programs worked on the project along with GIS specialists from ESRI's Redlands headquarters. The initiative was officially launched in February 2001 at the Global Ministerial Environment Forum held in Nairobi, Kenya, and has won widespread support as a means of integrating what was an immensely valuable but fragmented resource. It currently provides access to information on more than four hundred different variables, ranging from regional and global base maps to the distribution of greenhouse gas emissions, wildlife, protected and endangered sites, historical flood data, the condition of forests, and socioeconomic factors. Data sets are provided by more than thirty different organizations from around the world.

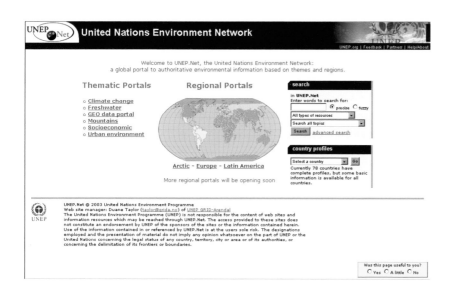

UNEP.net portal provides access to a vast array of data hosted by a number of different U.N. and related organizations. Data is organized around a number of topical or regional themes.

Many data sets are very rich, providing not just single maps, but also graphs, tables, and summary reports, all showing how variables change through time or space. Data is organized around a number of thematic (climate change, freshwater, mountains, socioeconomic) or regional (arctic, Africa, Europe) portals as well as a GeoPortal, which provides access to map and tabular data. The Geography Network technology is deeply embedded throughout the site, allowing reports, graphics, and textual data sets to be retrieved through spatial searches. The GeoPortal enables spatial data sets to be downloaded or analyzed on screen. As an example of its specificity, it is possible to generate thematic maps of national statistics on yearly fish catch for any year from 1960 through to 1999. Data can not only be mapped, but can also be represented in a range of dynamically generated graphs and tables, or downloaded in a variety of different formats.

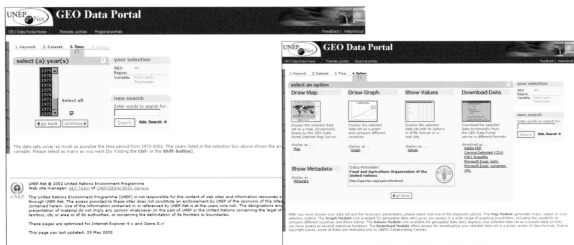

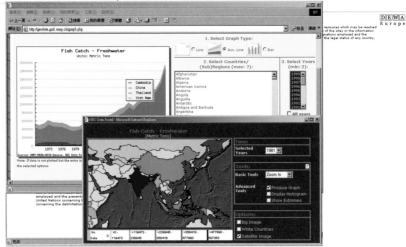

The UNEP.Net portal gives access to very detailed data sets. In this example, yearly freshwater fish catch statistics are mapped for 1981. The graph shows changes over time for selected Asian countries.

The UNEP.Net portal can be found on the Web at *www.unep.net*.

The UNEP.Net portal adds immense value to the existing network of environmental sites, complementing and enhancing them through consistent cataloging that permits data sets to be drawn together by shared location. The portal goes a long way to improving how data is disseminated to users around the world and helps raise awareness of the activities of participating organizations as well as awareness of the globe's current and anticipated environmental challenges.

National Mapping
Beyond paper to the Web

2

IT'S AN IRREFUTABLE TRUTH that the first requirement for running a national mapping office is paper—lots of it. Paper maps are the *sine qua non* of any national mapping operation, given tradition, and given the huge daily demand for printed maps, map series, and atlases.

This is, however, beginning to change as GIS Web services offer a more sophisticated digital cartographic environment that lets users design and download high quality, up-to-date maps of exactly the area they need when they need it.

This chapter looks at an example of just such a service from New Zealand, perhaps the first time anywhere in the world that a national mapping agency has begun distributing a new map series through a Web service, even before the printed versions are offered.

LAND INFORMATION NEW ZEALAND

Land Information New Zealand (LINZ) was established in 1996 with the reorganization of the Department of Survey and Land Information. Detailed topographic and cadastral survey in New Zealand began back in the mid-nineteenth century—more than one third of the land titles that exist today can be traced back to before the 1870s—and the entire country has been intensively surveyed. LINZ is responsible for custodianship of authoritative land and maritime data, supporting land ownership and transfer, topographic mapping, and the maintenance of standards for, and support to, the local surveying industry and for navigation.

Its maps and related products and services are used throughout national and local government as well as by surveyors, legal, commercial firms, police, and emergency services. LINZ also provides base mapping for environmental and park management in New Zealand's many national and maritime parks and its products are used by trekkers, mountaineers, and holiday makers from around the world, drawn to the pristine forests, mountains, and coastlines for which New Zealand is famous.

Since its inception, LINZ has been constantly updating both the quality of its mapping services and the way they have been delivered to its users.

Land Information New Zealand is responsible for providing New Zealand's authoritive land and seabed information. It provides a wide range of topographic, geodetic, and cadastral survey services. The organization is increasingly adopting the Internet as the principal method for information and services delivery, with TopoOnline a prominent feature of this initiative.

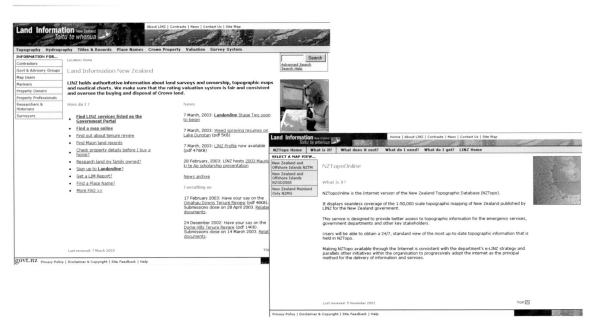

In the last five years this has included responding innovatively to New Zealand's e-government vision through a range of initiatives and product enhancements. Two of particular importance are the development of a completely new geodetic datum for national mapping, and the establishment of online service delivery channels.

Establishing a new datum for national mapping is a major undertaking. As a geodetic datum describes the mathematical framework on which accurate survey work and mapmaking is based, changing it means that all coordinates surveyed and maps drawn under the previous datum must be transformed to be consistent with the new one. With GIS and computerized data sets, transformation is a relatively simple process. But with vast quantities of map and surveyed material in general circulation, such a change, affecting all activity involved in producing or using maps, surveys, or geodetic coordinates, raises the potential for confusion—confusion that could have serious consequences for land title disputes, engineering works, safety, and search and rescue. As a result, national datums tend to be retained long after modern survey techniques have revealed inaccuracies within them.

Comprehensive civilian topographic mapping in New Zealand dates back to the 1940s, when a local datum (the New Zealand Geodetic Datum '49—NZGD49) was established. Based on this datum, a series of inch-to-the-mile maps (and derived small-scale maps) were produced in two Transverse Mercator projections—one for each of the two main islands. The metric products that followed this were derived from a 1:50,000 principal map series that was also based on the NZGD49. Initiatives to computerize the national cadastre and topographic databases that were undertaken in the 1980s and early 1990s both used the NZGD49 datum. However, as survey technology has developed, inconsistencies have been exposed within the NZGD49 datum. In addition, the NZGD49 could not be used directly with Global Positioning Systems (GPS) which are increasingly being used in the survey industry, the military, search and rescue organizations, and by recreational users.

After considerable study, a decision was taken to adopt a new datum (NZGD2000) that would provide a greatly improved accuracy, in line with modern survey techniques, and which would be compatible with international standards such as the World Geodetic System 1984 (WGS1984), and would enable direct use with GPS.

DATUM

▸ A *datum* is a numeric or geometric parameter on which other referencing systems or assumptions can be based. A geodetic datum is a parameter designed to produce a best-fit description of the shape and size of all or part of the earth's surface, and is the datum on which map projections and planar coordinate systems are based. As the earth is an irregular shape, any datum defined by a smooth ellipsoid will only ever be an approximation of the earth, and will depart from its actual shape in some locations. Many different geodetic datums have been developed to model a specific locale or country in which they are used or to model the entire earth. Accuracy varies among them based on the accuracy of initial geodetic observations and the assumptions made. Advances in recent geodetic instrumentation, particularly the use of satellite measuring systems, have improved our understanding and modeling of the earth and exposed limitations of a number of geodetic datums. LINZ adopted a geocentric geodetic datum, a geodetic datum that has the center of the earth as its origin—this makes mapping based on it easier to integrate with other data from other parts of the world, and enables direct use of satellite positioning systems.

GEODETIC DATUM

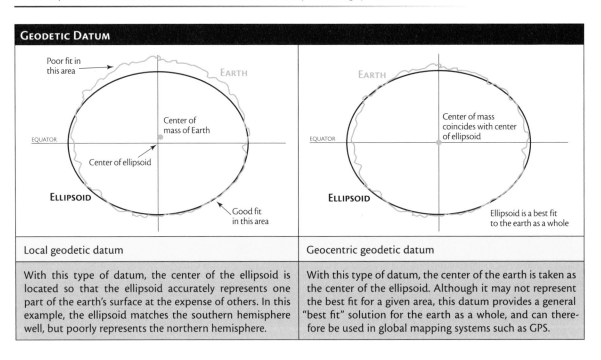

Local geodetic datum	Geocentric geodetic datum
With this type of datum, the center of the ellipsoid is located so that the ellipsoid accurately represents one part of the earth's surface at the expense of others. In this example, the ellipsoid matches the southern hemisphere well, but poorly represents the northern hemisphere.	With this type of datum, the center of the earth is taken as the center of the ellipsoid. Although it may not represent the best fit for a given area, this datum provides a general "best fit" solution for the earth as a whole, and can therefore be used in global mapping systems such as GPS.

Once a new datum is established, of course, a new map series must be created, and users must be made aware of the implications of the change. The implications were not trivial in New Zealand's case—the move to NZGD2000 meant all horizontal coordinates in the country shifted by around two-hundred meters. LINZ officials were also aware that they were not only serving users throughout New Zealand, but throughout the world, since New Zealand plays host to more than two million tourists every year, many of whom come for outdoor recreation such as trail hiking, climbing, biking, and canoeing. They rely heavily on LINZ's products and much georeferencing is based on the NZGD.

Having experimented with Web-based delivery of mapping since the late 1990s, LINZ was aware of its potential. They knew that it could reduce the time needed for updates to reach users, that it could ensure that different users could access the same version of a map, and that it could give new flexibility by allowing users to design and print their own maps. And, importantly, Web-based map services provided a focused Web application around which essential information on the impact and implications of the shift in datum could be disseminated.

In early 2002, LINZ therefore decided to launch an online mapping service, providing access to the new map series even before the corresponding hard-copy maps had been released.

FLEXIBLE, DETAILED, AND PERSONALIZED: THE NZTOPOONLINE MAP SERVICE

NZTopoOnline was officially launched in December 2002. It provides full access to both the new 1:50,000 series based on the NZGD2000 as well as to previous map series. It provides access to the very latest updates available to LINZ, ensuring that changes and updates are quickly available to map users. Targeted at the widest possible user base, it provides a full screen Web-based mapping service, and a simple and intuitive interface accessed through a standard Web browser. No plug-ins are necessary. A standard Web browser and a 28.8k modem connection are all that are required by the user.

NZTopoOnline provides public access to LINZ's new 1:50,000 base map series as well as to previous map series. The aim is to establish a system that permits very high cartographic quality maps in print-ready format, ready for downloading on demand.

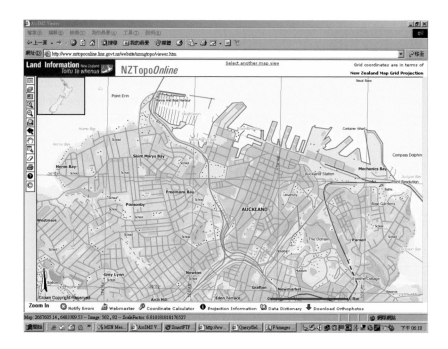

LINZ maintains a topographic database containing more than three-hundred clearly defined feature classes. Features to be presented on each map series are selected from this master feature set based on the scale and nature of the map and the content delivery method. The online service permits users to generate maps at almost any scale (although realistically this means anything between 1:5,000 and 1:1,000,000) and to add and remove layers to meet their particular requirements. Cartographic design, however, is a complex and highly skilled activity. One of the key considerations for LINZ was to let users access the flexibility offered by the new Web service, while recognizing that many would not have special cartographic or design training, and might need help to ensure the maps produced retained a reasonably high standard. As a result, LINZ worked on establishing a new set of rules for scale-dependent rendering that would enable meaningful maps to be generated at any scale. Viewed at 1:100,000 scale, basic topography, major roads, rivers, and first-level place names are displayed; at scales of 1:5,000, detailed contours (50 m intervals), street names, roads, streams, spot heights, and much local feature information appears.

NZTopoOnline can be found on the Web at *www.nztopoonline.linz.govt.nz*

NZTopoOnline provides access to more than one hundred data layers which are progressively revealed as the user zooms in on their area of interest. Controls allow layers to be reordered and turned on and off, permitting users to design their own map to be downloaded or integrated with their own systems.

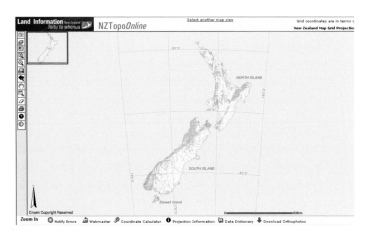

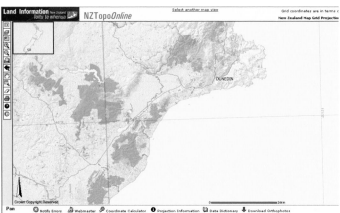

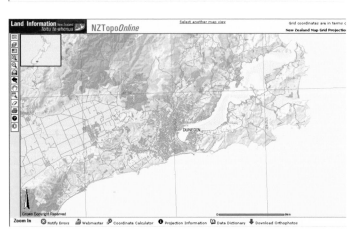

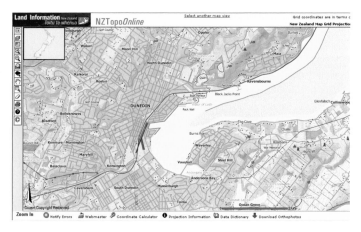

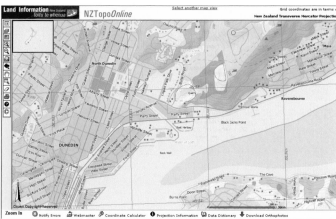

More than one hundred layers are available through NZTopoOnline, including a full range of topographic details, terrain models, and ortho-photos. Because the service conforms to OGC-compliant Web Map Server (WMS) standards, it is interoperable with other services complying with this standard. As a result, users who have data of their own in WMS-compliant systems (such as ArcIMS) can integrate this data and overlay it with their own layers. Thus, over the base topography, administrative data, or demographic information, market statistics, or the current location of a rescue team can be displayed. With the TopoOnline Web service users can, in effect, develop their own personal map service.

The features and symbology used in NZTopoOnline have been carefully selected from LINZ's core cartographic databases for optimal viewing at a scale of 1:50,000. A specially designed symbol set that includes more than 110 line, area, and point symbols is used to ensure that very high-quality maps can be rendered efficiently even on normal Internet connections.

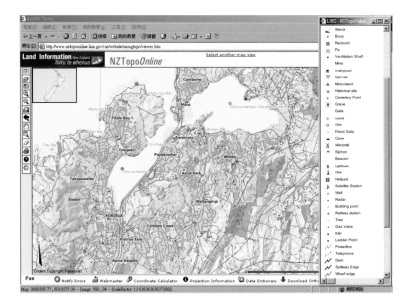

The system permits overlay of a variety of grids and graticules, including decimal degrees, making it easier to use data from GPS devices. Simple controls allow the user to pan and zoom the map and to select which data layers to view. There are tools to permit gazetteer place name searches and a road name search capability is also on the drawing board. Another key feature of the NZTopoOnline site are the links to a wealth of other services offered by LINZ, such as Web tools for coordinate conversion between NZGD49 and NZGD2000, data dictionaries and descriptions, download tools, and where appropriate, the capability to purchase and integrate vector data and orthophotos. Orthophotos at a scale of 1:25,000 (maximum pixel resolution of 2.5 m), are being progressively captured for the entire country and are available through the site in a variety of different image formats and resolutions.

The system encourages users to design and print their own maps that meet their particular requirements, whether these be route plans, regional store location plans, town maps, or mountain trekking plans.

WHAT'S BEHIND NZTOPOONLINE

NZTopoOnline is an ArcIMS 4.0 implementation. The configuration includes ArcSDE 8.2 on SQL Server 2000, serving one spatial server with four instance queues. The three map services are all HTML clients to ensure maximum browser compatibility.

The core LINZ data is held in a proprietary cartographic database. In order for data to be served through the Web Service it must be brought into a standard open format. This is undertaken with a routine maintenance system that keeps NZTopoOnline updated with changes made

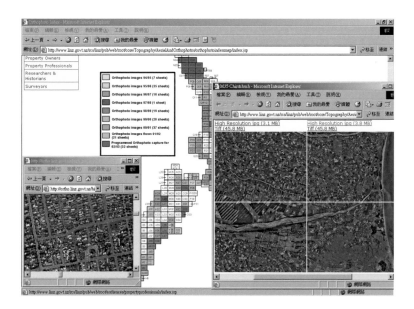

The system permits non-georeferenced orthophotos to be freely downloaded from the LINZ data library. Images are derived from annual surveys carried out since 1994, and will gradually cover the entire country. Areas where there has been rapid development will receive more frequent aerial surveys, and both current and historical images will be provided.

within the core LINZ database. Update routines extract changes from the core system as very large text files, and use automatic scripts to translate these to ESRI shapefiles and clean them.

Once the shapefiles have been optimized they are loaded into ArcSDE using administrative command scripts, again an unattended process. In total there are some 282 shapefiles loaded into the ArcSDE database, approximately 4 GB of vector data. The database is not versioned, so edits are always immediately committed.

The architecture of the system is a standard ArcIMS configuration. Data resides in SQL Server running on a dual Intel Pentium® III Xeon 1 GHz machine with 2 GB RAM and a 160 GB of RAID5 storage. It is accessed and maintained through SDE 8.2. The ArcIMS server runs on a dual Pentium III 1 GHz machine with a mirror backup to cover scheduled maintenance downtime and disaster recovery. The Internet connection is to the Wellington central business district's Citylink 100 MB fiber network.

NZTopoOnline is hosted within the DMZ of LINZ using dedicated ArcSDE, Web, and ArcIMS Servers connected to the Internet through a 10/100 MB switch. Off-site server and facilities are located at EAGLE Technology Ltd., who support and maintain the service.

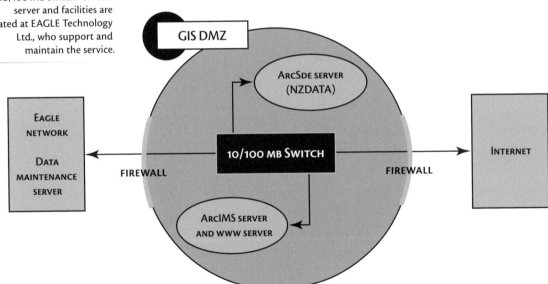

CONCLUSIONS AND FUTURE

NZTopoOnline is one example of the way LINZ has responded to the New Zealand e-government vision. LINZ is the custodian of data of great utility to many of New Zealand's citizens and visitors—the Internet and Web services offer an excellent way of disseminating this and expanding its use. In addition to NZTopoOnline, the department has also launched LandOnline, the world's first example of a completely integrated Web-based survey and land titles capture and processing system. Together these initiatives are moving LINZ's entire operations, from data capture to eventual service delivery to the end user, into a computerized, Web-enabled environment. These systems have also helped smooth the intro-duction of the NZGD2000 datum and the dissemination of information about it.

These projects are attracting the attention of other national mapping agencies around the world who are interested in replicating them in their own countries. LINZ has been an earlier mover. This in part testifies to the foresight of the department, but is also a result of the national gov-ernment's policy of permitting, indeed promoting, free and easy access to publicly held topographic mapping. Elsewhere, national map agencies are often forced to charge money for large-scale topographic map data, or, for security reasons, to actively restrict publication and distribution. The ability to integrate e-commerce engines with GIS Web services (as is shown in a number of the examples discussed in this book) and increased security of Internet-based transmission may move to overcome some of these obstacles. NZTopoOnline, however, acts as a demonstration of what is now feasible.

In the near future, NZTopoOnline Web Service will be enhanced to enable the download of very high-cartographic-quality maps in print-ready formats—real "on-demand" printing that can be accessed from anywhere through the Internet. This will be the last step in removing the requirement to hold large stocks of paper maps and an extended hard-copy distribution network. In practice it is unlikely that map sheets and print runs will disappear from national mapping offices in the near future. However, as NZTopoOnline shows, those days may well be numbered.

Hosted Map Services for Government

3

Arguments for GIS implementation in local and national government are well understood and widely accepted. GIS provides a common geographic framework that enhances data sharing and integration; builds bridges across departmental or discipline boundaries; and facilitates reporting and presentation throughout the organization and to the public. But establishing and maintaining such systems can be complex and time-consuming, and can require significant investment in data, systems, and staff resources. GIS services accessible across the Web and hosted by outside contractors offer a cost-effective alternative.

This chapter examines how this has been accomplished successfully by several organizations in the United Kingdom, where local and national organizations have found that ESRI (UK) Ltd.'s MapsDirect hosted service can provide them with the kind of rapid deployment of GIS services and efficient support for their standard data and mapping requirements that they are looking for.

MapsDirect Central Government enables multiple organizations to access gigabytes of Ordnance Survey base mapping from a single, professionally managed hosted service. The simple interface provides generic visualization and map navigation tools. User organizations can either customize this basic system, or can integrate the service directly with their own Web and desktop mapping systems.

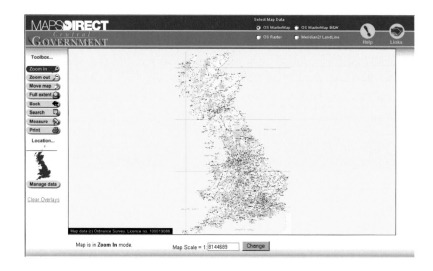

Although the U.K. government has been making use of GIS for many years, access to and management of data had become a significant issue for many central and local government departments. The U.K.'s national mapping agency, the Ordnance Survey (OS), provides some of the best digital mapping in the world. Although itself a government agency, the Ordnance Survey operates as a self-financing entity, covering all its operating costs through the sale of products, services, and copyright licenses. Private companies and government entities both pay fees for the use of OS data.

Although they were by necessity one of the major users of OS data, government organizations were finding that both the cost and the overhead required to manage complex data licensing agreements had become a significant burden for them, particularly in the smaller departments. The situation was resolved, however, with an agreement between central government and OS reached in April 2002. The agreement established a unified, single licensing regime that permitted data to be used in more than 550 government departments. It greatly simplifies the licensing and use of OS digital map products, opening the opportunity for GIS and mapping to be placed on desktops throughout all levels of U.K. government.

Licensing issues resolved, the challenge for government organizations then became a more technical one—how to take maximum advantage of this data by providing it in a simple, easy-to-use manner, one that would facilitate access and use by as many users as possible. This was, and is, no easy task. Modern data sets from the Ordnance Survey are both large and complex; the latest OS MasterMap series is vast, with each layer containing millions of features, and, when stored in Oracle and accessed using ArcSDE, some 450 gigabytes. It is also dynamic; more than five thousand changes per day can be made to data in the core data sets. Serving this quantity of data to users across a large organization and keeping it up-to-date requires a major investment in hardware, software, networks, system development, and maintenance skills. This effort is duplicated in every single department or agency maintaining its own copy of this data set. MapsDirect was established to provide a simple hosted service that provides reliable access to key OS data sets, can be accessed by multiple organizations at the same time, and provides generic, easy-to-use, flexible spatial functionality.

MapsDirect

GIS departments in large organizations such as local and national governments are busy places, continually updating and maintaining data; developing, rolling out, and maintaining corporate mapping interfaces; looking after hardware, software, and networks; managing data licenses and copyrights; and developing custom applications and analyses. Increasingly, as mapping is integrated with mission-critical services and disaster response systems, all this must be provided twenty-four hours a day.

Based around the same set of hosted textual and spatial databases, a family of MapsDirect generic packages is being produced to cater to the needs of particular user groups.

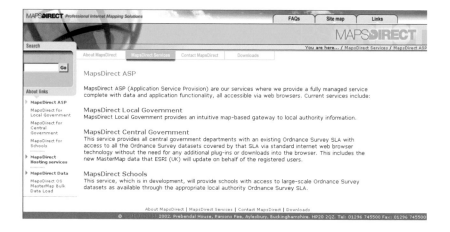

MapsDirect is a fully hosted ArcIMS-based Web service that provides a variety of services to both public and commercial organizations, including dedicated services to local and national government bodies. The aim is simple: to provide as a single license package a flexible mapping tool that delivers up-to-date, actively maintained map data that can be used throughout an organization and can be integrated with existing applications and data sets. Providing these services to multiple clients brings real economies of scale and enables the kind of active data maintenance and investment in hardware, application development, and disaster recovery facilities that are often beyond the abilities of all but the largest organizations.

MapsDirect is on the Web at *www.maps-direct.co.uk.*

MapsDirect services are hosted within a secure environment in a three-tier architecture separating Internet, application, and data servers. Key servers are mirrored to provide downtime resilience. OS base maps are hosted on a single central data server that is accessed by all services regardless of the client to which they are provided. The data server is maintained on a machine with 636 GB RAID5 disk array and 2 GB of RAM. Data is managed in Oracle 9™ and ArcSDE 8.1. Services utilization is monitored to provide load balancing during peak periods and ensure dedicated allocation of bandwidth to particular sites and services.

Maps are served in ArcXML format. A suite of standard Web mapping tools is provided which can be used as the basis for rapid development of a new mapping interface. Alternatively, MapsDirect can serve data directly to existing Web and desktop application clients.

By taking over the mundane work of licensing and data update, and providing core mapping facilities, MapsDirect permits an organization's internal GIS staff to concentrate on their primary roles—maintenance and development of that organization's own data resources, and development of organization-specific analyses and applications.

CROWN ESTATE: NATIONWIDE MAPPING FOR A NATIONWIDE ESTATE

One of the first to adopt the MapsDirect system was the Crown Estate, the agency responsible for managing the diverse land, property, and marine possessions that are part of the hereditary estates of the British Crown. This portfolio spans the length and breadth of the British Isles and includes property scattered across major towns and cities, some 300,000 acres of agricultural and rural land, more than half of the foreshore (land between high- and low-water marks), and rights to the seabed stretching out to the twelve-mile territorial limit. Though technically owned by the Crown, these properties are actively managed to generate revenue that is passed through to the national treasury.

Crown Estate, responsible for one of the largest land and marine estates in the United Kingdom, was an early subscriber to the MapsDirect service.

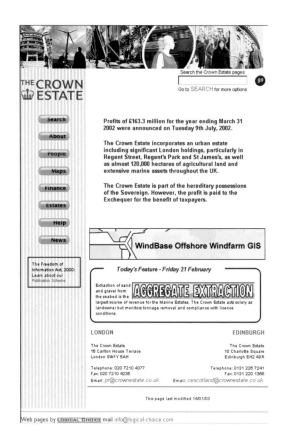

Managing a real-estate portfolio that spans the nation requires high-quality mapping that also spans the nation. Until the Ordnance Survey's pan-government license came into effect, Crown Estate operated with a collection of hard-copy maps and a number of independent GIS applications covering specific assets. The pan-government license agreement permitted unlimited access to national mapping (including the new MasterMap series) to be provided to all Crown Estate users. Rather than establishing a new application to disseminate these map resources across the organization, or modifying existing independent applications to take advantage of it, Crown Estate decided to use MapsDirect for Central Government—a specifically tailored MapsDirect interface for government agencies—which effectively delivered both solutions while keeping disruption and implementation costs to a minimum.

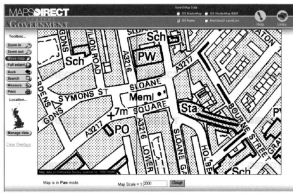

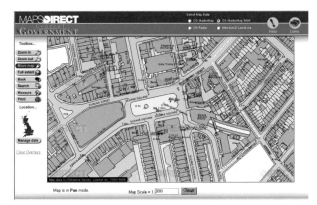

MapsDirect for Central Government provides complete U.K. coverage for a comprehensive set of both raster and vector Ordnance Survey map products. These are updated as Ordnance Survey releases new data. A simple ArcIMS-based interface provides tools to allow easy map navigation, layer configuration, and high-quality printing—all within a standard Internet browser that can be accessed through all computers on the organization's network. Geospatial Web services permit gazetteer searches based on postal code, street address, geographic coordinates, or OS Tile reference. The interface permits Crown Estate data to be integrated and overlaid within the standard Web-based interface, but it also permits MapsDirect base maps to be directly served to existing desktop applications—ensuring that all users, whether using the light browser-based interface or an existing application, can access the same, up-to-date, nationwide data.

The service hosts a range of Ordnance Survey products. Shown here are OS MasterMap, Raster, and LandLine products, focused on a neighborhood in the Chelsea district of London. As soon as updates are released by OS, the hosted services are updated. The generic viewing services provide navigation and scale-sensitive rendering.

Finding a location anywhere in the United Kingdom is simple with the MapsDirect service, and can be done through a range of search windows or interactively with the cursor on screen.

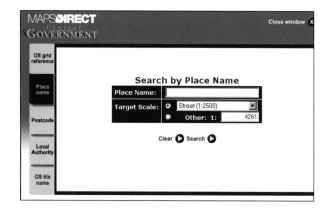

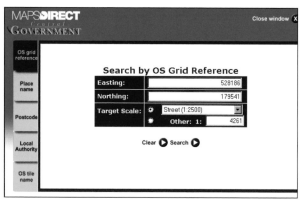

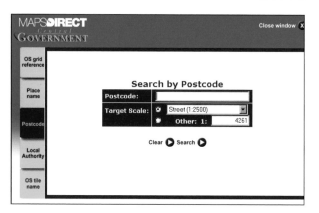

Wiltshire and Swindon Pathfinder: A multi-agency GIS portal

The Wiltshire and Swindon Pathfinder (WASP) system links a range of county-based government bodies in the first multi-authority GIS portal in the United Kingdom. Participants in the project included Swindon Borough Council, Wiltshire County Council, and the District Councils covering West and North Wiltshire, Kennet, and Salisbury districts as well as Wiltshire Health Authority, police, and fire services.

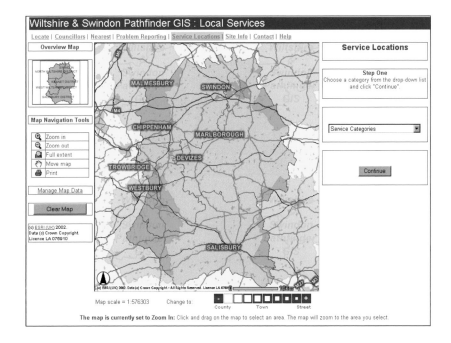

Covering a large area of central southern England, the Wiltshire and Swindon Pathfinder GIS Web service provides an integrated service bringing together data and functionality from a number of county and district councils in the area.

Developed as part of the U.K. government's Pathfinder initiative to encourage local e-government service delivery, WASP provides a single point of contact for citizens and local government workers alike to search and locate services and facilities provided in a geographic area. Developed and launched within less than two months, WASP represents a powerful demonstration of how hosted Web services such as MapsDirect can cut implementation times and simplify delivery of core mapping functions.

MapsDirect serves up a range of Ordnance Survey data sets that form the base map for the system. License watermarking and copyright notice inclusion on prints and screen captures are automatically handled by the system. On top of this a wide range of data sets covering local services, amenities, and points of interest are drawn from existing systems (often held in incompatible formats) maintained by the participating organizations.

WASP permits users to locate services by name, type, address, or proximity. They can find their local government representative, school, library, or the community center that serves their particular area. All data is geocoded and is presented both graphically on the base map, and in a series of textual or graphic information screens drawn from related databases. The system also facilitates accurate citizen reporting, allowing users to identify the exact location of graffiti, blocked drains, faulty road lighting, fallen trees, or similar problems.

The WASP service enables users to locate services and facilities anywhere in the region, search for services in their neighborhoods, find out who the local representative is for a particular area, and report problems. In this example, the WASP Web service has returned all schools within a 2.5 kilometer radius of a user-defined point (the small cross). Their locations are displayed on the map pane and their address details, and exact distances from the point are displayed in a report window.

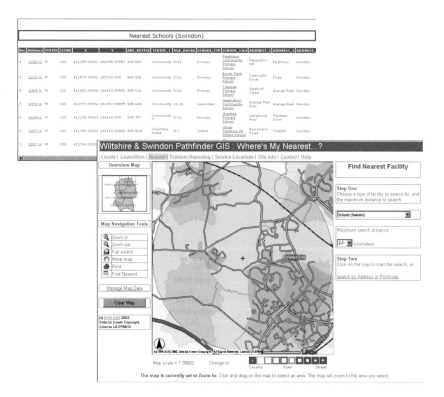

Future enhancements are planned: new data sets and interfaces for new district Web sites, and the expansion of the system to provide base mapping to existing desktop applications.

CHICHESTER DISTRICT COUNCIL:
HOSTED WEB SERVICES GIVE ROOM TO EXPERIMENT

Chichester District Council is using the MapsDirect hosted Web service on a trial basis to gauge public demand for information, and service utilization parameters. The council has established aggressive goals for electronic delivery of services, including placing more and more information and services on the Internet. One key goal was the launch of a map-based information portal to enable citizens to search and locate key facilities and services. Rather than take the risk of developing such a portal from scratch, Mark Jennings, the GIS analyst with Chichester's e-Government Team decided to implement a beta version based on MapsDirect. By monitoring traffic on the trial system and the reaction of the users, detailed plans could be drawn up for the eventual development and launch of an internally hosted service.

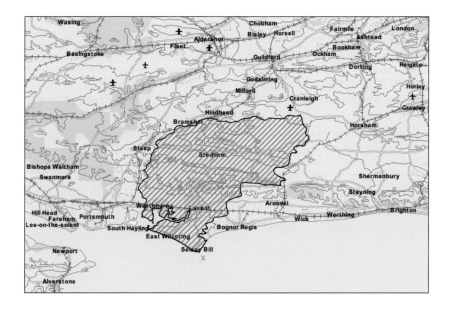

Chichester District Council is located in southern England and covers just over 303 square miles (approximately 486 square kilometers). The area is primarily rural and, containing many historic villages and towns, is a popular tourist destination. One of the incentives for adopting the MapsDirect service was to assist visitors to the area as well as local residents.

The interface for Chichester District Council MapsDirect looks radically different from those described previously but all are powered by the same hosted data and GIS search services.

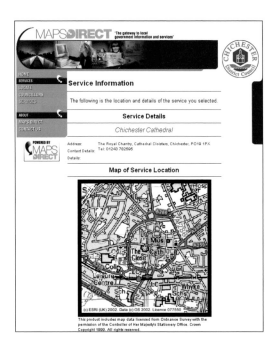

"The advantages of this approach for Chichester were clear. MapsDirect hosted service provided us with a cost-effective, quick, and easy way of deploying an application to test the market and see if our citizens found such access useful. Basically all we have had to do is to provide location-based data sets containing location, address, telephone numbers, and so on, and then to decide on the background mapping that would be used."

— Mark Jennings
GIS analyst with
Chichester District Council

Built on top of standard MapsDirect hosted services, the Chichester system provides Ordnance Survey base mapping and access to standard location-based Web services that are integrated with local data. Thus, the user is provided with options to do location, point-of-interest, and proximity searches. They can also find out who their local representative is, and search for council facilities and services in their area. With implementation time measured in terms of days rather than months, and the ability to record detailed monitoring statistics, MapsDirect is providing the council with exactly the kind of test application they were looking for.

HOSTED WEB MAPPING SERVICES

Web services such as MapsDirect have a compelling logic for local and national government organizations. Local and national governments all use standard base data sets and require fairly well-defined search, visualization, and printing tools. Allowing the development and maintenance costs of both the interface and the ongoing update and support of base map data sets to be shared across a number of organizations is both efficient and cost-effective. As Web services evolve, such services are being more and more tightly integrated with existing applications to the extent that the user is largely unaware of the source location of the data or functions used, or where the maps are being generated. Such developments help government organizations not only improve their internal information networks, but also to improve how they interact with their public and how they disseminate information.

Enterprisewide Data Integration and Visualization

4

MANAGING DATA is an active, dynamic, constantly evolving task. Certainly it involves ensuring that data resources are securely held, updated, and maintained in good order. But fundamental to responsible data management is ensuring that data resources actually get used—only then can the value of the data resource be realized and the expense of its collection, storage, and update be justified. The task for the data manager thus includes actively promoting access to data sets, and identifying new ways of disseminating, integrating, and presenting them, so that maximum benefit can be reaped from their collection.

For public organizations, from national governments down to the smallest district or town council, which are often the custodians of vast quantities of public data, this mandate means ensuring that data is put to work for the benefit of the communities they serve. For private organizations it means keeping data actively deployed for commercial or economic advantage. For both public and private organizations, achieving this goal involves promoting flexible and easy access to data sets throughout the organization to ensure that decisions based on these data sets are thoroughly informed and coordinated. It can also involve providing data to potential users outside the organization either as chargeable or freely available data services. To achieve this coordination, integration and presentation of data sets is critical.

In practice, the task of data management is not easy. The volume of data handled, its diversity, the frequency of updates, the number of different departments using it, and the range of the interests those departments serve all conspire to make responsible data management a very major undertaking. In addition, the way data is used varies depending on the user and circumstances—it is almost impossible for the data custodian to foresee all possible uses for a particular data set.

GIS Web services are beginning to provide a solution to this challenge. Used in conjunction with Internet portals, they are enabling data to be linked and integrated in new, highly flexible ways, and to be disseminated easily to a wide range of users both inside and outside of the data-holding organization. This chapter looks at how Web-based GIS services in the form of enterprise information portals are being deployed in two organizations in Australia to enhance enterprisewide integration of, and access to, their data resources.

CITY OF GREATER GEELONG

The City of Greater Geelong can be found on the Web at *www.geelongcity.vic.gov.au.*

Located on Port Philip Bay at the southern tip of Australia, the city of Geelong is the state of Victoria's largest provincial city and one of Australia's leading industrial centers. Covering an area of around 1,300 square kilometers, it encompasses a major port, a mixture of heavy and light industry, low- to medium-density residential development, and farmland. The area is also famous for its pristine beaches facing the Bass Strait and Tasmania, and for a variety of important ecological sites, including sensitive wetlands, and state and national parks and sanctuaries. With 1,800 employees, the City of Greater Geelong is Victoria's largest local government.

The city council was created through the amalgamation of several smaller local governments in 1993. Working hard to consolidate data management within the new organization, by the late 1990s managers had organized data into a set of key core corporate systems that were maintained and used by different departments. But the city did not stop there. Integration of the data held within these core systems was the city's goal. A further challenge was the need to incorporate a wide range of data sets owned and managed by external organizations and data service providers. Data, regardless of its source, had to be presented in

a simple, easily accessible manner available to all employees at any location, through a single, corporatewide interface that would not require
extensive training. An enterprisewide Web-based GIS presented a suitable solution.

The diverse data sets typical of a city administration included property
and customer information, planning data, environmental records, city
infrastructure such as road and drainage assets, utility supply details
such as water and sewer mains, waste and recycling collection zones and
bins, public assets such as parks, community, and recreational facilities,
transport, emergency, and regulatory services including animal, building,
retail, and public-health licensing records. The most obvious common
connection among these data sets was shared location, and council staff
found that one of the most accessible ways to work with and access this
data was through a map-based interface. If such a map interface were
Web based, any solution could also be distributed across the corporate
intranet, ensuring that the service could be reached from anywhere
within the organization. In 2000, Geelong therefore commenced the
implementation of an EView enterprise GIS Web service.

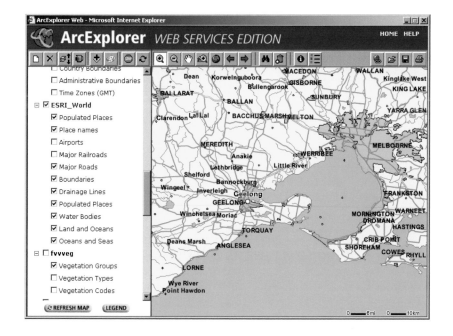

Geelong covers an area of
1,300 square kilometers, and is
home to 190,000 residents.

EView is based around the establishment of a central spatial database that acts as the integrating hub connecting data sets held throughout the organization. A lightweight Web browser application provides the access and analysis functionality to users on the council's intranet. The central spatial database, known as PLACES (Planning, Land Assets, Community, Environment, Services), was established in an Oracle RDBMS accessed through ArcSDE 8. Other textual and document databases continue to be maintained by individual departments utilizing their existing applications, and are linked to the PLACES database either by address or by geographic coordinates. Data—spatial, textual, and image data sets—are presented to most users through the browser as a single unified interface for interaction with all data sets. This greatly expands access to core corporate data—permitting it to be displayed and analyzed from all of the council's 650 personal computers. The Web browser interface is highly

The Eview Web service brings together data sets and functions from across the council enabling them to be integrated, viewed, and worked with from every single computer on the council's intranet.

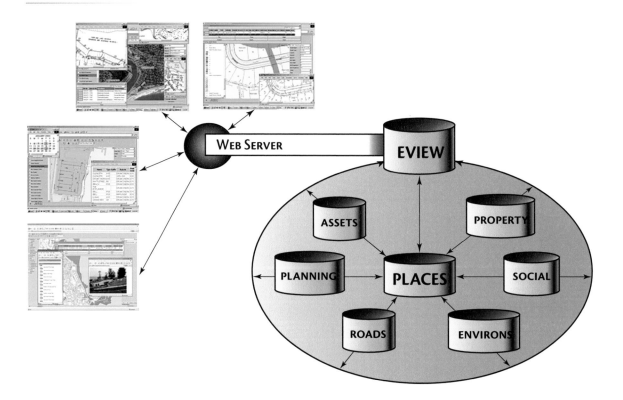

intuitive and the majority of council staff members can use it productively after only a couple of hours of training.

The coordinated approach to data management has also allowed it to be streamlined. The system is managed with a very small core team. Two staff members, a system administrator and a cartographer, are assigned to the system full time and are responsible for maintenance and update of the spatial layers and the linkages between them and corporate tabular databases. A further eleven staff members distributed throughout different departments and offices have been trained to use ArcGIS 8 and act as power users, not necessarily working on the system full time, but able to undertake complex analysis or data maintenance tasks if required, and acting as points of contact, giving advice and assistance on the Web-based interface when they are needed.

MAXIMIZING USE, MINIMIZING EFFORT

The map interface provides access to more than 150 of the council's map layers and acts as the vehicle for accessing other textual and graphical data sets held throughout the organization. The system brings together data to support all aspects of the council operations with one-stop access to textual, spatial, and document information on properties, customers, licenses, animals, services availability, waste collection, assets, social infrastructure, biodiversity, and the environment.

Users navigate the map interface either by typing a key such as an address, owner, ward name, or asset identifier, or simply by progressively zooming in on the area of interest with a cursor. The familiar map navigation zoom and pan tools are available, and layers can be turned on or off and their order changed as needed. Related information can be retrieved from council databases using a simple identify tool and clicking on the feature required, or through a flexible report-building interface. The system provides rapid, easy-to-understand graphic answers to the large number of everyday customer service requests such as *where is my drainage connection point? Or what services are available? Or what are the names and addresses of adjoining owners for fencing? Or what planning or environmental controls will affect my building application?*

For a given area, Eview can return a vast range of information including textual reports, schedules, statistics, images, CAD drawings, and of course, maps.

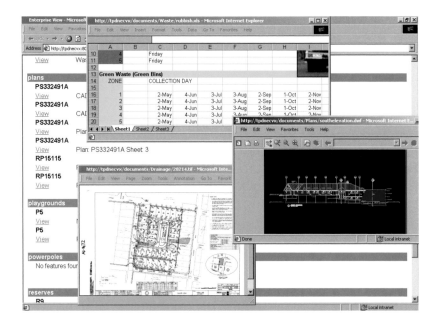

EView keeps track of individual user sessions, permitting step-by-step queries to be built up as a user explores various data sets in the course of retrieving information. This enables users to follow a logical workflow, progressively narrowing down a search by taking the results of one query as the starting point for the next operation. It extends the strength of the system to allow flexible and powerful combinations of inquiry.

This can be seen in the way it simplifies a typical city operation—road maintenance. Based on customer complaints regarding the surface of a particular road, EView can access the Asset Management system and identify the extent, condition, and maintenance schedule for the related road segments. The user can then select all properties affected by the proposed maintenance by buffering the interconnected road segments. Then, by accessing the Planning database, the user can refine this selection to include only residential, commercial, or industrial holdings. The community and emergency services layers can also be rapidly checked for any potential critical impact. From this, reports can be created for letters of notification to be mailed to the property owners, businesses, and affected services. Thus, in a single sequence of operations, the user has traversed the mapping, asset management, planning, property, emergency and community infrastructure databases—each held and maintained in discrete departmental systems—analyzed the impact of the work and generated the necessary notification without having to leave the desk or knowing the schemas of the databases used.

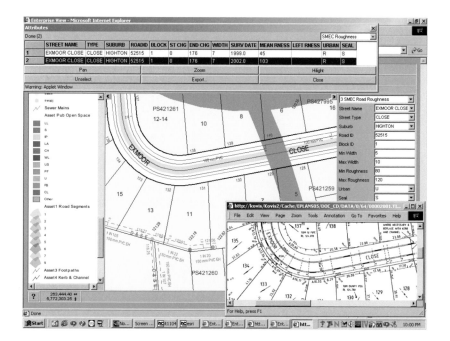

Eview provides access to the asset database maintained by the roads department including multiple layers of information on road layout, condition, street facilities, and related plans and reports.

Running on the council's intranet, the EView portal is accessible on all of the council's 650 computers distributed across several offices. Though at present the service is provided for in-house users only, both the server and browser are entirely configurable through XML, and could provide similar services through the Internet to the public or third-party entities. E-business solutions are a practical possibility that is being explored by the city for the future.

The system has not only greatly increased exposure and use of the council's data resources, it has also had a major impact on management and training overheads. The browser-based interface completely removes the need for application configuration and maintenance on the client side. Moreover, because of its simplicity, staff training and software license costs are kept to a minimum.

The implementation of this service has meant that data resources held by the council are available to a far wider group of users and are, as a result, being more frequently referenced. By providing access throughout the council and simplifying many of the data integration and management tasks, investment in data and information systems can be maximized.

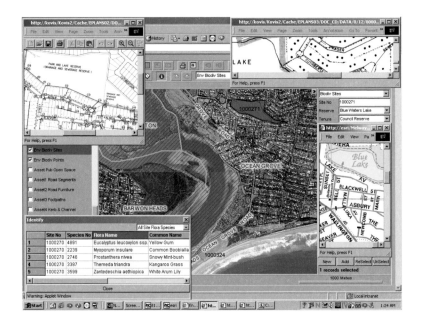

DEPARTMENT OF LAND ADMINISTRATION, WESTERN AUSTRALIA.

The Department of Land Administration, Western Australia can be found on the Web at *www.dola.wa.gov.au.*

The Department of Land Administration, Western Australia (DOLA) has adopted a similar approach to managing and distributing its data resources. Western Australia encompasses 2.5 million square kilometers, and is arguably one of the world's largest states. DOLA has responsibility for the production of hard-copy and digital mapping throughout the state, as well as for land title registration and valuation and for management of Crown lands, which account for 93 percent of Western Australia's total land area. As the state's primary surveying and land-management agency, the department maintains a comprehensive listing of land, survey, and cadastral information.

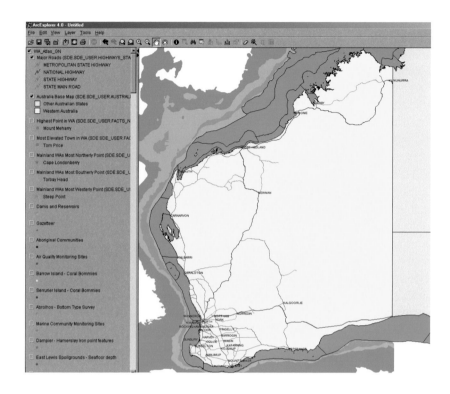

Covering a massive area—2.5 million square kilometers—Western Australia covers a third of the Australian continent. DOLA Western Australia has its main offices in Midland, and other offices throughout the state.

Traditionally, obtaining data from DOLA has been accomplished by mail or by traveling to one of the three department offices throughout the state—a time-consuming and inefficient process both for those requesting the search and for DOLA staff. However, since 2000 DOLA has been providing an increasing amount of data and services through online systems. As part of its commitment to digital service delivery, DOLA has recently launched LandLinks, an online service built on EView that provides effective, integrated access to land and survey data.

The underlying objective of the LandLinks service is to:

▹ provide access to high-quality land and spatial information
▹ achieve a fair return on the government's land and property information assets
▹ streamline processes supporting the Western Australian land market
▹ advance the Western Australian land information industry

The LandLinks service consolidates data held in five separate databases (cadastre, tenure registry, topography, native title, and geodetic framework) maintained by different sections of DOLA. Departmental policy requires that no direct public access is permitted to these core corporate databases. Data, both textual and spatial, is therefore transferred nightly from the secure departmental environment to the LandLinks server and updates both ArcSDE layers and Oracle databases. The Landlinks server is only used for disseminating data to the public. This arrangement not only ensures that the corporate network remains secure, it reduces the chance of ongoing work or unverified data being accidentally released into the public domain.

The service is designed for professional surveyors and land agents and is only available to registered users. It allows them to interact with data sets, to design and produce maps and reports, and to download related information across the Internet. One of the key benefits of EView is that unlike many Web-based applications, the state of each user's activities is maintained at the server end, avoiding the use of cookies or other client-end techniques. Maintaining the user state not only permits sustained searches and analysis to be carried out in a natural, multistep manner, but it also provides significant performance savings.

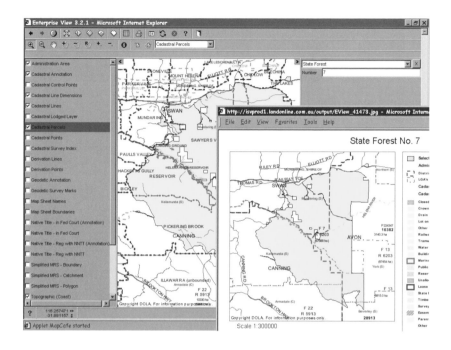

The DOLA's LandLinks service enables users to access the vast data resources held at DOLA. The service permits data to be retrieved through both textual and spatial queries, viewed and interrogated on-screen. A print service enables the user to generate customized maps of selected data, which, alternatively, can be downloaded in a variety of formats.

The service is currently structured around three basic views.

The General View provides basic cadastral and topographic information including base mapping with topographic features and high-level planning zones, and land tenure data sets. Land ownership can be viewed and selected based on a number of different criteria including address, lot, leaseholder, or whether they are state or Crown lands, statutory environmental reserves, or parks.

The Survey View provides the land survey industry with access to DOLA's detailed cadastral and geodetic information including cadastral angles, points, control points, survey indexes, derivation information, and geodetic marks. Data is both mapped and provided in tabular format where appropriate. In addition, historic information can be obtained because the portal also provides access to scanned images of the original Survey Index Plans.

The system enables surveyors to identify geodetic control points in the areas where they are working and to download location plans and station description data that can be loaded directly into their applications or instruments.

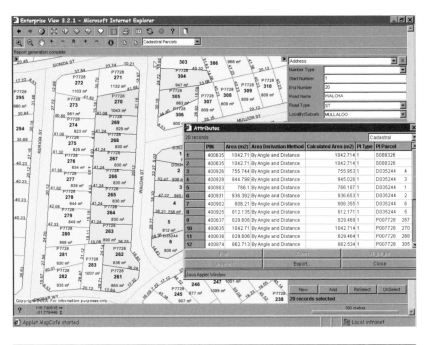

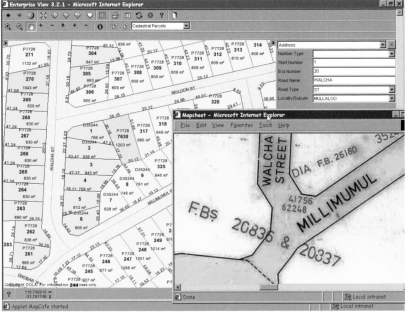

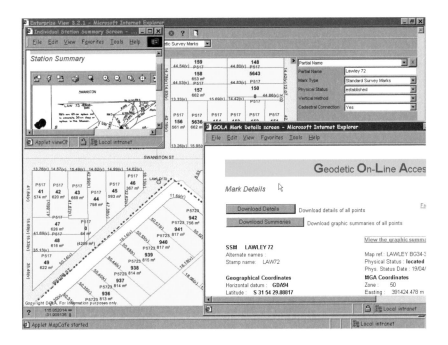

The Native Title View is specifically designed for those involved in Native Title land claims and provides access to the details and history of lodged claims, presenting these in both graphic and textual format.

Provided across the Internet and requiring only a Web browser to operate, LandLinks WA has greatly expanded the ability of the department to reach and support its customers, and to ensure that the data it holds is put to use. Users have access to a consolidated set of departmental data presented in an integrated, coherent manner. Searches can be conducted at any time and from the comfort of users' offices. In addition, the system is encouraging more proactive feedback from users. As it is now easier to access digital data and to visualize this against other data sets, users familiar with the local area can also more easily identify errors or areas that need to be updated. Feedback options within the interface allow users to alert the department to areas of concern and these can be accompanied by sketch plans identifying the suggested changes and other documentation. Any change will still require full investigation and measurement by DOLA, but having an efficient feedback channel to inform the department of issues within the data sets greatly facilitates database maintenance.

DOLA DATABASES

- **Topography:** forms the fundamental base map showing natural and manmade features, hydrography, vegetation, and relief, and height. The scale at which the data is captured varies between 1:2,000 in urban areas and up to 1:100,000 for areas of desert and rural areas.

- **Cadastre:** containing the legal boundaries of every block of land in Western Australia including: housing lots, parks, forests, farms, pastoral leases, reserves, and roads. Land parcels are digitized from the largest scale base mapping available at the time of capture. Each parcel is assigned a centroid with a unique parcel identifier number which links it with the tenure register.

- **Tenure register:** containing ownership of land parcels, including parcel identifier number, owner's details, plot address, certificate of title, surveyed area, purchase price, date of sale, and land use.

- **Native titles:** details of native land claims submitted to the federal court or the National Native Title Tribunal including location, details on those lodging the claim, current status, and key dates and progress through the legal process.

- **Geodetic framework:** including details of both current and removed benchmarks used for horizontal and vertical survey. Details include base coordinate information to various datums, accuracy statements, and notes.

At present, the system is provided as a subscription service, but opportunities to provide new and more flexible services in response to growing user demand are being explored. Currently these include transaction fees, and direct integration with systems and data from other government agencies, or from community or private-sector organizations.

Rather than having disparate systems and services providing customers with access to specific segments of this vast collection of information, DOLA is looking to develop a one-stop shop for WA government's land and property information. This could be customized and adapted for a client's particular business needs through the adoption of Web- based components such as EView.

ENTERPRISE VIEW: DATA INTEGRATION MADE EASY

A geographic hub, managed through ArcSDE, is always found at the center of the EView architecture. Maps and map features (held in one or more ArcSDE databases) comprise the nodes to which diverse databases are tied and through which corporate information resources are discovered. While effort is required to establish initial linkages, many of these can occur automatically with geographic functions and based on existing coordinate or address files.

ArcIMS is used as the Internet map server. The browser and server applications are based on Web standards including XML, HTML, ArcXML, and Java. Though a standard browser application is provided to facilitate rapid deployment, this is highly flexible and can be customized or completely replaced if required. The system can easily be integrated with desktop clients (such as ArcView or ArcInfo™) through its Java programming interface and, with plans for deployment of SOAP connectors, other third-party software applications conforming with this standard. The overall architecture provides a flexible, scalable environment through which corporate data can be exposed as a variety of different Web services depending on the nature of the audience.

For more information on EView refer to *www.esriau.com.au/products/pages/internetsolutions/10835.*

Streamlining Property and Planning Through the Web

5

FOR MANY, buying a home, or building a new one completely from scratch, or even just adding a room to our current residence is the fulfillment of a long-held dream. But realizing that dream can be a long and frustrating business. For buyers, finding the right home and having an offer accepted is only half the battle—after that come the lawyers and surveyors and the seemingly endless round of property checks and title searches to ensure that all is in order before the deal is finally sealed. For those planning to build from scratch, or to add an extra bedroom, the planning review process provides the equivalent test of patience and wills. Few would question that such processes are essential to protect the public and the environment where they live—it's just that they take so long.

Web services are bringing together technology and data to smooth out and shorten this traditionally arduous process. This chapter looks at two examples of initiatives that make use of map-based Web services from the United Kingdom: the national NLIS Searchflow, and a local planning system that has been established by Sevenoaks District Council, in Kent in southeast England.

National Land Information Service (NLIS) is a joint initiative by both national and local governments in the United Kingdom that aims to speed up the delivery of publicly held land and property information used in legal searches. Designed as an online one-stop shop, the system permits search requests, and in some cases dynamic information retrieval, from data sets held by multiple agencies throughout the country. At the local level, district councils such as Sevenoaks are increasingly turning to the Internet as a means of keeping both residents and applicants informed of the progress of particular applications and local planning activities.

NLIS SEARCHFLOW: WEB SERVICES SIMPLIFYING CONVEYANCING

When a property is bought or sold, an often laborious verification process must follow. Does the seller have the right to sell? Is the definition of the boundary of the property valid? Is the property's attractiveness, structural safety, or value likely to be affected by factors that may not be immediately obvious to the buyer? Will it be, for example, affected by existing planning restrictions, or future development plans nearby? Or perhaps there has been previous industrial activity that has undermined the foundation, or contaminated the soil or water sources? In the United Kingdom this verification, or checking process, called conveyancing, has traditionally been undertaken by lawyers or (more recently) property agents working on behalf of the buyer. Because the information needed to answer each of these questions is held by many different organizations, each check traditionally required a separate inquiry, often by mail, to a different organization and incurred, of course, a separate fee. The result is that conveyancing has often been both time-consuming and expensive.

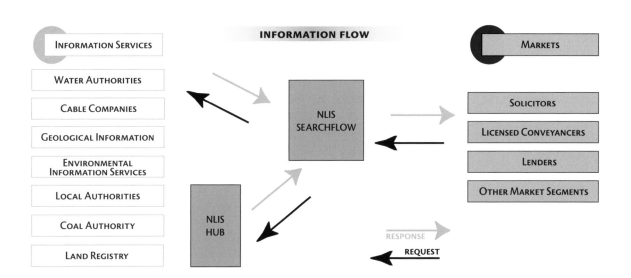

NLIS Searchflow is a Web service that provides a gateway through which the majority of these checks can be processed. It aims to achieve three fundamental goals:

▸ To provide a means of obtaining and validating an unambiguous definition of land or property;

▸ To identify all relevant information sources that may hold data relevant to a particular transaction;

▸ To provide an efficient means of ordering, tracking, and/or extracting information from these information sources.

NLIS Searchflow does this by linking a number of discrete online services hosted by different information providers to create a single point of access through which searches can be defined and orders dispatched. Information on land title and property boundaries from the Land Registry, local planning conditions from local government, and certain industrial or mining activities that may affect the area (through subsidence or soil contamination, for example) from organizations such as the national Coal Authority, are all accessed through the national government's central NLIS hub. In addition, NLIS Searchflow permits access to the information services provided by a number of other public and commercial organizations, including water and cable companies, and specialist organizations holding additional information on a variety of factors

NLIS Searchflow can be found on the Web at *www.searchflow.co.uk.*

including industrial activities, environmental hazards, flooding, transportation, and geology. Together these provide all the basic information required for a full property search: land title, property boundaries, previous transaction history, known industrial activities in the area, local environmental hazards, current or planned developments in the area, planning restrictions, and similar factors.

The sources used are, more often than not, identical to those involved in a traditional paper-based search. However, the search request, and all necessary information to define its scope and nature, can now be electronically submitted to every relevant data-holding organization in a single operation. Some of the data providers have hosted their own online database resources from which information can be pulled directly, while others allow requests to be logged and validated electronically—though the search itself remains manual. Either way, for the lawyer or agent issuing the search requests, it is a significant improvement, greatly increasing efficiency and reducing costs.

Because the entire service is Web-based, integrating a number of remote services, there is also no need for the user to buy or maintain additional hardware, software, or data. Once subscribed to the service, the system can be accessed from any standard Microsoft Internet Explorer v.5 or higher browser.

The system enables properties in question to be located either by typing in an address or name, through identifying it on a map, or with coordinates. In the left column, a monitor keeps track of progress and indicates the next step to be undertaken.

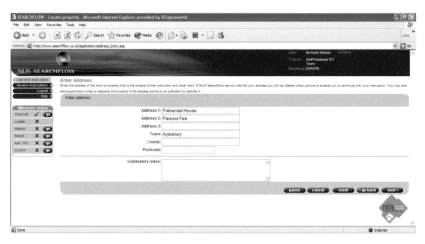

The address or partial address having been entered, the system will search an up-to-date U.K. address database and return possible matches from which the user can choose.

NLIS Searchflow helps users navigate the often treacherous legal waters of property conveyancing, through the use of an easy-to-follow, step-by-step process.

An NLIS Searchflow search starts by defining the property in question. Inspection of the most recent base maps for the area is an essential first step for all searches because this will highlight features such as adjoining property, shared pathways, or easements that may impact property ownership and will determine the nature of the search. Traditionally this has required ordering base maps covering the property in question from the Ordnance Survey, U.K.'s national mapping agency. While in time lawyers may build up a reasonable set of large-scale base maps for their local area, because they can be dealing with property anywhere in the country, and because base maps are regularly updated and therefore change, inevitably they need to be ordered specifically for each search. Using NLIS Searchflow, the lawyer or agent is immediately presented with a centrally maintained up-to-date base map covering the entire country. If the property's address or postcode is known, this can be used to locate and center the map display on the property in question. Where there is no defined address (say in an rural land transaction), the researcher can use map navigation tools to zoom in on the area of interest. Either way, an up-to-date map of the property can be retrieved in seconds rather than days.

Based on an analysis of the map presented, the user can define a search area with a cursor that may, depending on the nature of the surrounding features, be considerably wider than the property itself. Once the search area is entered, NLIS Searchflow starts to create a draft search request—completing and validating address details, generating search area coordinates, and producing a search definition map which will accompany the request. Maps are generated automatically by the remote map service.

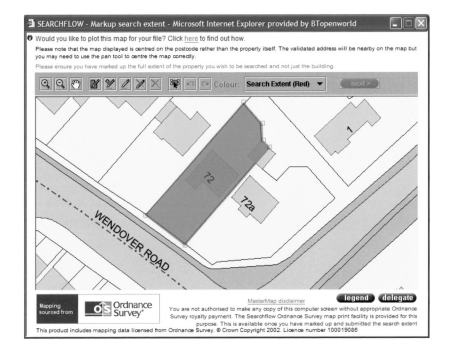

An example of the map service embedded within NLIS. Users can search and select property with the cursor. The service accesses an up-to-date OS MasterPlan base map, ensuring that users can check existing road and boundary details. If, as is often the case, a document search needs to include features of the surrounding area, such as footpaths, adjoining walls, or boundaries, the user can interactively define the area to be included in the search with a cursor. The map image and defined search area can be saved and printed and are automatically attached to any search request issued from the system.

The system then uses the defined search area to establish a comprehensive list of information sources that are relevant to the site. Drawn from NLIS databases and from a range of different local and national organizations, this list-creation function greatly simplifies the task of the conveyancer, who frequently may not have detailed knowledge of the area where the transaction is taking place. To facilitate selection, the system prioritizes those searches deemed to be most relevant, but ultimately the lawyer is responsible for confirming or amending this list. Based on the selected list, the system will then automatically calculate the total search fee—summing all the search charges for individual data providers as well as royalty and license fees for the maps generated. NLIS Searchflow is linked

Based on the user-defined area of interest, the system automatically identifies all searches that are relevant for the particular property in question and allows the user to select those that they wish to conduct.

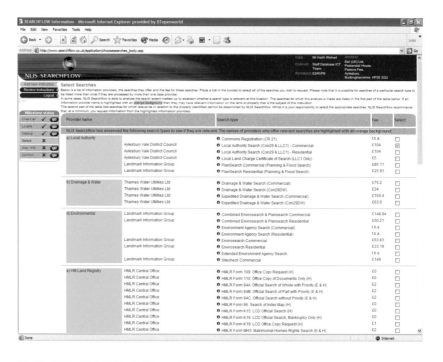

It is possible to refine any given search to specify exactly what details are required. On completing the selection, all search request documents are compiled automatically, including the user's details, property address, and map location plan, and are sent to each of the individual authorities or organizations that will process the searches.

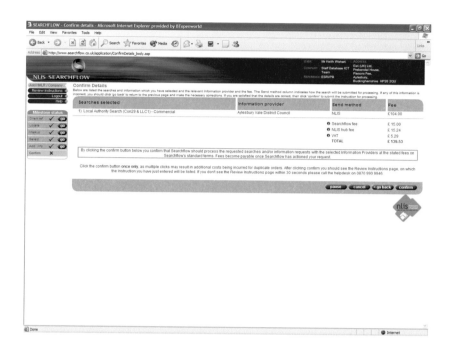

Search fees payable to the organizations involved, along with processing charges, Value Added Tax, and any royalty fees, are automatically calculated, and compiled into a single invoice.

with an e-commerce engine which handles all payments to participating organizations. So the lawyer issuing the search no longer needs to pay each individual organization separately—the system handles that, and the lawyer makes a single payment.

Once payment has been confirmed, requests are disseminated to the relevant information providers. Some providers have established services that take NLIS Searchflow requests as input and can process the response automatically. Others can receive NLIS Searchflow requests, but because the searches still require manual processing, they return the information either as an e-mail attachment or through the post. Because the system helps accurately define which searches are required and automatically generates and verifies many of the entries in forms submitted (for example, ensuring postcodes and street addresses match), it is greatly improving efficiency for both the lawyers and the information providers.

Winner of a number of legal, e-commerce, and government awards (including the Department of Trade and Industry's Information Society Initiative in 2000, the Ecommerce Award in 2002, and the Society for Computers & Law award, also in 2002), NLIS Searchflow is making a significant impact on the speed and cost of searches required in buying a house. Figures suggest that on average, searches are processed five to six times faster than with traditional techniques, with a similar magnitude of cost savings.

SERVING PLANNING INFORMATION WITH PUBLIC*ACCESS*

Planning is an equally intensive process. It requires the circulation of planning applications to a host of different departments and organizations—including architectural, environmental, legal, water and electricity, and fire services—as well as soliciting the views and opinions of the general public. The results of these evaluations must be tracked, coordinated, and finally consolidated so that a balanced, justifiable decision can be made. All of this must be undertaken within a relatively short eight-week timeframe, the target set by U.K. government for responses to be given to planning applications. For busy planning authorities that can receive several hundred applications per month—ranging from multistory office and retail complexes complete with new parking and road layouts, to simple attic extensions—this is a major undertaking.

The Sevenoaks Public *Access* Web site is 195.217.202.176/publicaccess.

Sevenoaks District borders the southeast edge of London. Only a half-hour train ride from central London, it is within the city's prime commuter belt. A stone's throw from London's two large international airports and the Channel tunnel rail link to continental Europe, it is also a popular choice for business and light industry. With the whole of southeast England continuing to experience dynamic growth both in economy and population, pressure on land for housing, for amenities and recreation, or for retail and industrial uses, is intense. Some of the busiest planning offices in the country are found in the district councils in this area, and planning issues are followed closely by local residents and businessmen alike. Sevenoaks is no exception, receiving the highest number of applications in the county of Kent.

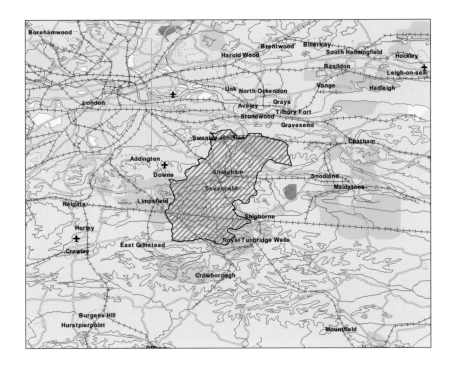

Sevenoaks is located just to the southeast of Greater London and covers a total area of 142 square miles (228 square kilometers). In prime commuter land and close to two of London's international airports, Sevenoaks District Council has one of the busiest planning offices in the area.

In order to speed up processing applications, since 1998 Sevenoaks District Council has used the CAPS Solutions's UNI-*form*™ system to manage and track all aspects of planning. UNI-*form* is a single integrated database environment maintained by different council departments involved in the planning process to present a comprehensive view of application status. The system is based on an Oracle database storing spatial data in ArcSDE to provide access to relevant maps covering the area of planning and planning applications.

In 2002 the Council introduced Public*Access* which is built on ArcIMS and which brings UNI-*form* data to the Internet, allowing the general public to view details, track progress, and comment on planning applications online.

Public*Access* serves two functions: it permits users to track progress and decisions of a given planning application, and it gathers comments on current applications. It enables members of the public to query live planning application information held within the Development Control module of the UNI-*form*. Information presented includes: applicant name and address, type and nature of application, key dates, current status, case officer, and related graphics or plan data in scanned format. Planning data is dynamically retrieved, and comments, objections, and other entries from interested parties are directly input into the system being used by planning officers. Clearly, since users need to orient themselves in order to use the system, a map interface must also be provided.

The Public *Access* service allows users to select an address or location, and to check planning activity in the area. Here, a search is undertaken on a particular street. If a planning application has been received that affects that street, its location will be displayed on an up-to-date base map (yellow dot). In addition, the service summarizes application history, giving key dates, brief details of the proposed development, current status, and links to additional information.

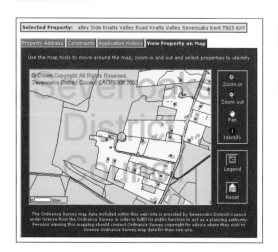

Sevenoaks District Council was keen to implement Public*Access* as soon as it became available, but their Internet hosting infrastructure was in the process of being upgraded at that particular point. As a result, the delivery of Public*Access* was initially externally hosted—a task undertaken by ESRI (UK) Ltd. using its MapsDirect servers (see chapter 4 on MapsDirect). This meant breaking the link between Public*Access* and the live UNI-*form* data; however, a satisfactory solution was established to allow routine weekly feeds of UNI-*form* data to the externally hosted site, and then the immediate transfer of public comments and objections entered through Public*Access* back to Sevenoak's UNI-*form* system. With the infrastructure now in place, the council is progressively transferring back responsibility for hosting and maintaining the system.

The site permits access to planning information in a number of ways. Users who know the relevant planning application number can enter it, and check site plans and status directly. Because members of the public not directly involved in an application are unlikely to know the exact application number, the system facilities execution of quick searches by grouping applications by submission week, or by allowing the council to flag "Special Interest" applications that may have particular interest to the local community. Once a property has been selected, maps of the area can be retrieved along with details of the application, current status, relevant dates, and any comments received from the public.

It is also possible to check whether there are any pending or previous applications relating to any property in Sevenoaks. Using the cursor to click a building will bring up a list of previously submitted applications. It is possible to review these, and, for those that have already been decided, to receive a summary of the decision and any constraints or conditions that were attached to it.

ESRI (UK) Ltd. can be found on the Web at *www.esriuk.com.*

Not only is this process helping to keep applicants informed of the status of their applications, the system facilitates gathering public comment on planning issues and particular cases, and provides easy access to previous decisions made. The latter is particularly valuable for those considering submitting an application in the same geographic area, because it means they can design their proposals with this awareness, ultimately leading to better prepared and more targeted proposals. While the Public*Access* Web service does not eliminate the need for scrutiny of planning applications by a wide range of council departments as well as by the general public, it does facilitate the process, and it ensures that applicants and the general public alike are better informed about the planning process and about the specific details of applications in the area.

Mapping Volatility

6

ASIA, home to some of the world's most volatile property markets, has seen prices tumble from the exorbitant heights of the late 1980s and early '90s, but there is little doubt they will skyrocket again. Even in a slump, property values remain some of the highest in the world, and with China, Indonesia, India, and Pakistan boasting some of the world's fastest urban and economic growth rates, the boom will probably come sooner rather than later. That certainty—that things will remain uncertain—makes valuation, or appraisal, of property portfolios in the region a challenging, but essential, aspect of financial strategy for the individual, for corporations, and for government.

Success in this demanding sector depends on a host of factors—local knowledge, timing, technical expertise in a wide range of disciplines, including economic and structural forecasting, engineering, architecture, design, and of course, location. Location means spatial analysis, a discipline that plays a critical role the majority of property valuations.

Despite its importance, maintaining spatial data sets and establishing a corporate GIS platform has remained a seemingly daunting task for many companies involved in valuation, one that few have attempted. But one busy Hong Kong valuation office that has ventured into the GIS space has found that GIS Web services offer an alternative that saves time and money. It puts targeted spatial functionality across the corporate intranet, at the fingertips of everyone in the enterprise.

KEEPING A WATCHFUL EYE

With three hundred staff, operating across China and Southeast Asia, Chesterton Petty Ltd. is one of Asia's largest property consultancies. Part of the Chesterton International Group, with offices around the world and a track record in the business of almost two hundred years, the company has been involved in the region for thirty years, and in some of its most significant development projects, private and public. Its Valuation Group, based in Hong Kong, works throughout the region on valuation projects; including hotels and resorts; government and corporate land sales; and commercial, industrial, and residential developments. Keeping clients up to date with the latest fluctuations in property markets in such places as Shanghai, Bangkok, and Hong Kong is a major challenge.

Asia-Pacific's leading provider of
PROPERTY SERVICES

Chesterton
PETTY
卓德

Chesterton Petty provides an integrated property consultancy and agency service. Our aim is to work in partnership with our clients, delivering solutions that combine an in-depth understanding of their needs with our specialized knowledge and expertise.

Member of the
Chesterton Binswanger group

160 Offices Worldwide
• Asia • Australia • Europe
• The Americas • Middle East • Africa

| Market Reports |
| News Desk |
| Directors |
| Contact List |
| Client List |

| Services |
| Housing Channel |
| Homes International |
| Blue Solutions |
| About Us |

| Worldwide Offices |
| Jobs |
| Contact Us |
| Feedback |
| 中文主頁 |

| Special Property For Sale & Lease | | Commercial Property Listing |

Chesterton Petty Ltd

 For Lease

Chesterton
PETTY

Market Reports
News Desk
Directors
Contact List
Client List
Services
Housing Channel
Homes International
Blue Solutions
About Us
Worldwide Offices
Jobs
Contact Us
Feedback

< Home >   < Index of Services >

Two International Finance Centre

Hong Kong Station, Central

Prime Location
Prestige
Convenience
High Efficiency

MTR Central Station
MTR Airport Express Hong KongStation
Airport In-town check-in counter

The tallest building in Hong Kong - 88 storeys

The highest office floors available for lease:
20,000 - 23,000 sq.ft.lettable per floor
Efficiency: 93% approx.
Extra Features:
- Dedicated lift lobby
- 100% back up electricity supply for all sockets

False ceiling height:
2.7m (general floor)
3.3m (trading floor)

Raised floor system
150mm (general floor)
300mm (trading floor)

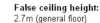

With thirty years' experience in Asia's most dynamic property markets, Chesterton Petty Ltd. is one of the largest property consultancies in the region. Accurate appraisal has been the key to their success.

Meeting that challenge requires that the company build and maintain a network of sources and local knowledge, and build experience with a number of diverse disciplines and with the cities themselves. It also requires strong backup support to ensure that databases and records are maintained and accessible; that reports produced in different offices and regions are coordinated and consistent; and that deliverables reach clients in a timely, efficient manner. As part of its effort to maintain and continually improve its technical support systems, Chesterton Petty Ltd. began exploring the possibility of providing a Web-based mapping service that would enable spatial analysis and map production to be accessed from across the entire organization.

Maps are used at every stage of the Valuation Group's work. When a new project comes in, maps are referenced to get an accurate location and to identify any other projects in the area that may have been completed recently. In the initial stages of an investigation, a number of map sources may be referenced to establish the value of the property. For a residential development assessment this might include checking zoning plans, neighborhood descriptions (with criteria such as age demographics, income levels, or crime patterns), and proximity to desirable features such as historical or cultural points of interest, public facilities, and schools. For industrial and commercial properties, assessment is likely to include reference to land tax zones, proximity to potential clients and transportation hubs, and the potential for expansion. For larger projects that may span a number of years, maps can be used to collate and index information on the study area. In addition, presentations to clients typically include detailed location plans and maps of the environment, supplemented in larger projects by thematic maps illustrating particular issues.

While Chesterton Petty Ltd. used ArcInfo GIS software for a number of particularly large, complicated projects, it was not integrated with the day-to-day workflow. For many projects the company still relied on a variety of hard-copy street guides, atlases, and sheet maps, and the traditional techniques of photocopy, overlay, and sketch and paste.

GIS FOR CONSOLIDATING INFORMATION

With almost twenty offices throughout Asia, and four in Hong Kong alone, maps were frequently—and unnecessarily—duplicated among offices. Keeping these up-to-date and consistent was a constant struggle. Internet mapping aroused interest within the company, because it offered a way to develop a central GIS map database that could be accessed from anywhere on the corporate intranet. But the in-house IT team had limited familiarity with Internet GIS and equally limited experience hosting spatial data across a distributed network, least of all one that covered a large geographic region such as Southeast Asia. In any case, the IT team had plenty of work to do supporting other necessary services. A demonstration by ESRI China (Hong Kong) Limited of ESRI's Geography Network in operation in mid-2002 led to the beginnings of an idea of a hosted Web service mapping engine. Further discussions between Chesterton Petty Ltd., ESRI China (Hong Kong) Limited and potentially interested data providers resulted in basic requirement and design specifications being drawn up by fall 2002.

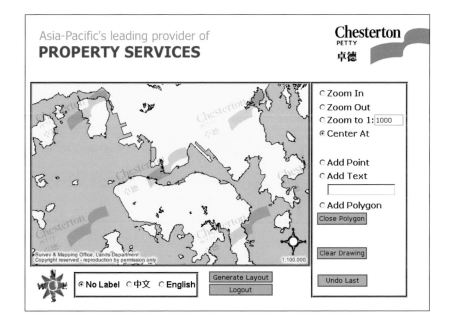

The interactive Web mapping system enables Chesterton Petty Ltd. to provide up-to-date, consistent map data throughout their offices. Generic Web services have been combined to form a customized interface providing map navigation, layer control, search, and basic map drafting. All data and interface services are drawn from external Web service providers, greatly minimizing development costs and ongoing data and system management overheads.

The interactive Web mapping system was established as a customized Web service hosted on ESRI China (Hong Kong) Limited's Geography Network. One of the advantages of this approach was the speed of development. Chesterton Petty Ltd.'s mapping engine was implemented and deployed within two weeks of system specification. Though the interface is tailored specifically to meet the needs of the Valuation Group, the majority of functionality behind it comes from integrating a number of individual GIS Web services that were already hosted on the Hong Kong Geography Network. These include such generic spatial applications as:

▷ Access authentication—providing login and usage tracking services;

▷ Map display—layer and symbol control as well as map rendering;

▷ Map navigation—zoom, pan, and scale-dependent layer selection;

▷ Place name search—permitting searches by street address, area name, or point of interest;

▷ Geocoding—enabling point features to be generated from addresses or from a variety of coordinate systems;

▷ Spatial searches—based on proximity;

▷ Measurement tools—location, distance, and area measurements;

▷ Output—production of printer-ready output.

The interactive Web mapping system is a custom solution hosted on ESRI China (Hong Kong) Limited's Web service architecture. The same architecture and basic server components can be served to multiple clients in different ways. The Chesterton Petty Ltd. solution uses a SOAP interface, and three of the available Web services.

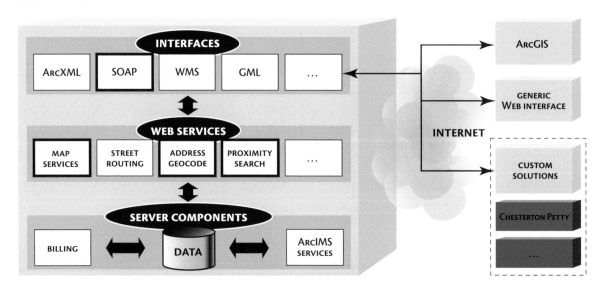

By being compliant with the XML standard, consolidating these individual services to form a new customized Web service application was straightforward. As they have already been built, tested, and tuned, system development and testing were extremely fast.

Accessed across the Internet, the system is available to any member of staff regardless of location. Problems of maintaining hard-copy maps in multiple offices and consistency in the map products output have disappeared, since all staff can access the same standard base map. Maintenance and update of data is handled as part of the hosting service by ESRI China (Hong Kong) Limited.

A COMPREHENSIVE DESKTOP VALUATION TOOL

Available on users' desktops, the system also saves time spent in the library or searching for relevant map sheets. For example, to do a valuation of a new residential development, the user can type in an address and the map interface will automatically center itself on the property in question. Or, if the address is not available (in the case of a proposed development or one under construction) the user moves the cursor to zoom to the area in question. Neighborhood features can be verified because map layers showing nearby facilities such as stations, bus stops, parkland, and shops, or planning zones and socioeconomic data, can be displayed. A measure–distance tool can then be used to define and measure walk paths between the property and nearby facilities. Alternatively, a simple buffer tool can be used to undertake straight-line distance searches; integration of a new service that will enable searches to be based on a navigable road network is also planned. Additional layers, for example those showing previous projects undertaken by Chesterton Petty Ltd., are periodically updated to the hosted service, so that users can easily check reference file numbers of any previous work that has been done in the area. In effect, almost all operations that used to require hard-copy maps can now be undertaken by staff without leaving their seats.

If a location plan is required, the system provides a simple HTML output template that allows the property or area in question to be highlighted, and the map annotated. Standard legends, scale bars, grids and north arrows, and titles can be added if required. Alternatively, the map

window can be captured and exported directly to the company's desktop publishing applications. In either case, the system automatically inserts any standard copyright watermarks required by the data set providers.

The interactive Web mapping system is developed using Microsoft .NET framework, which enables rapid development of applications consuming Web services. Microsoft .NET automatically handles many of the communication protocols, including authentication access and secure delivery through Secure Socket Layer (SSL) connection. The hosted Web services are developed in ArcIMS 4; they access data sets managed in an Oracle 8 database through ArcSDE 8.2. The service runs on a series of server-class machines with built-in redundancy to allow twenty-four-hour availability and maintenance requirements.

The map service uses map styling so that data from a single data set can be streamed to clients in different ways. This enables the user to switch easily between annotating the map in English or Chinese without having to open another map pane. As the Chesterton Petty Ltd. service expands to cover other countries, this will be used to establish similar bilingual interfaces.

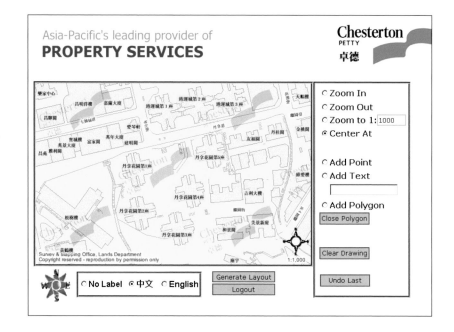

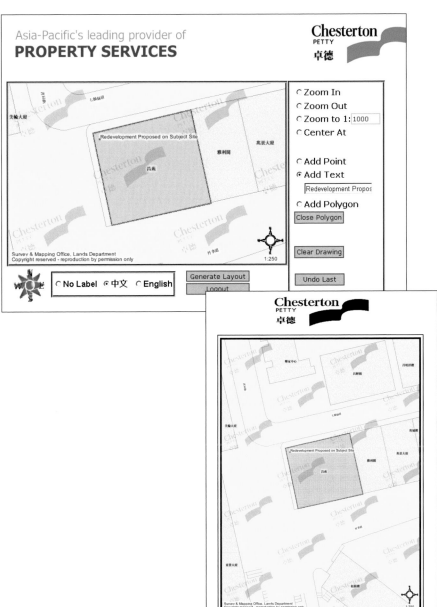

Simple location plans can be generated, highlighting the property in question and allowing annotation text, symbology, and titling to be included. This can be output to a simple HTML template for printing or inclusion in reports.

CONCLUSION

The first and most obvious benefit of the system for Chesterton Petty Ltd. is the ability to ensure access to up-to-date, consistent map resources across the company. This is improving the efficiency and overall coordination of location analysis and map production in the company. Although location and mapping remains only one of many factors that are involved in valuation, the Web service approach enables this aspect of the overall process to be addressed in a very sophisticated way, while minimizing in-house development and maintenance costs. At present, the system covers Hong Kong, and expansion is planned for other key cities where Chesterton Petty Ltd. operates, including Shanghai, Beijing, and Guangzhou in China, as well as Singapore, Manila, and Seoul.

Monitoring What Moves Where with Web Services

7

THE ABILITY TO MONITOR OBJECTS on the move—boats, cars, people, herds of elephants—has become so commonplace that it seems difficult to imagine that there was once a time when a boat could weigh anchor and its location would remain unknown for months. Now, small, lightweight, and affordable global positioning systems (GPS) can report current location to an accuracy of ten meters. Even without GPS, the location of a mobile phone user can be pinpointed to within fifty meters. Tracking systems are now firmly established as a viable commercial service for businesses and individuals. For business, they offer the ability to improve both logistical and customer-response efficiency; to monitor fleet movements and operational patterns; and to improve safety. In the consumer marketplace they have proven themselves by enabling location information to be easily reported in an emergency, or by allowing the tracking of vulnerable members of society, such as the young, the disabled, and the elderly.

GIS Web services are building on these technologies to host tracking services for users across the Internet and, for those who are themselves on the move, through wireless connections.

Trackwell can be found on the Web at *www.trackwell.com*.

This chapter focuses on the Icelandic company Trackwell, which has been developing tracking technology from its beginnings in traditional vessel tracking applications through to modern commercial Web service tools. Its story illustrates the close relationship between the expansion of the tracking and location-based services industries with the Web services industry, and also the way that Web services can be used to migrate existing applications to new platforms.

VESSEL MONITORING ORIGINS

Based in the island country of Iceland, it is perhaps not surprising that Trackwell's first involvement in the technology focused on the movements of ships in the surrounding waters of the Atlantic. The ocean is central to Iceland's development and economy—its fishing industry alone accounts for more than 70 percent of its export earnings and provides employment for more than 10 percent of the workforce. The waters around Iceland are some of the richest fishing grounds in the world, essential not just for the local fishing industry, but for those of Europe and the northeastern coastal communities of the United States and Canada; the Icelandic fishing areas attract vessels from as far away as Japan and Korea.

Trackwell's initial systems were designed to improve safety at sea. Search-and-rescue teams routinely dispatch in response to distress calls. The ability to provide accurate tracking of vessel movements before and during an emergency incident saves lives not only of those onboard vessels that have got into difficulties, but of the rescuers that were sent out to help them. Trackwell's first systems were established for the Icelandic Life Saving Association and for Iceland Telecom, based on research conducted by University of Iceland's Engineering Research Institute. The Icelandic Coastguard became an early adopter of the system and helped fine-tune its specification. That initial system permitted vessels to report their location to coast-guard headquarters either automatically, using a variety of satellite communication systems, such as Inmarsat or EMSAT, a shore-based VHF system, or manually, by using voice radio, telex, or fax. The system recorded vessel locations into a database and presented them on a simple GIS platform.

However, it quickly became clear that this technology could be expanded to help monitor and manage all fishing activity around Iceland. A series of international agreements and treaties agreed through the North Atlantic Fisheries Organization (NAFO) and North–East Atlantic Fisheries Commission (NEAFC) provided an international framework for the management of north Atlantic fish stocks. These organizations represent the collective interests of countries with fishing fleets working in the North Atlantic, and their agreements on zonal or seasonal restrictions, catch quotas, and other measures are designed to ensure the maintenance of healthy fish stocks. For these organizations, monitoring the location and status of fishing vessels would help surveillance and enforcement of agreements.

ICELANDIC FISHING SURVEILLANCE

▷ Initial vessel tracking systems were established after an incident in 1967. Hit by a sudden violent storm, a local fishing vessel sank before crew members could send a distress call. Although all crew members got into life rafts, they stayed there for five days in rough seas because as no distress call had been sent, no one was looking for them. The next year, Iceland established the Icelandic Vessel Reporting system, which permitted fishermen to radio in their positions upon departure and return to port, and every morning and evening while at sea. If a boat failed to make contact, rescue teams were alerted. The Icelandic Life Saving Association initiated trials of the first automatic system in 1988.

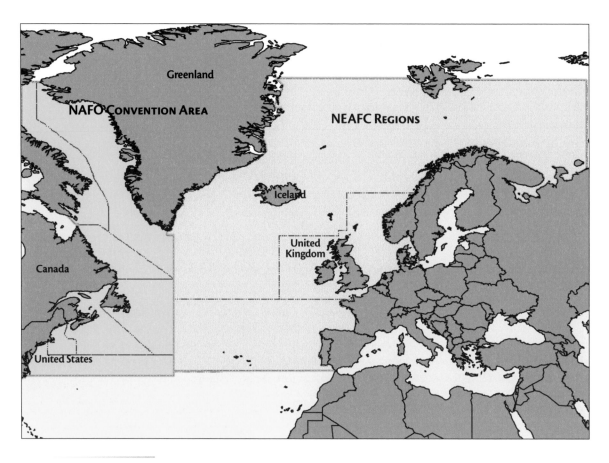

The NAFO and NEAFC Convention areas cover the whole of the North Atlantic. Within these broad boundaries there are many subareas and blocks defined which form the basis for monitoring and reporting the health of fish stocks and fishing activities.

The initial tracking application developed for the Icelandic Coastguard was enhanced to meet specifications defined by NEAFC, which were later adopted by NAFO. These specifications required expanding both functionality and data storage. The Vessel Monitoring System (VMS) maintains basic fleet information, including size and type of vessels, ownership details, size of crew, and the kind of fishing the vessel engages in. It permits vessels working in the area to submit routing itineraries, catch and status reports, and automatically record current position. Vessel location, track, speed, and direction bearing, along with relevant zone boundaries and navigation chart details can be displayed on an online map. Current position is automatically compared with agreements and exclusion zones defined by NEAFC. This means that if a vessel strays into such an area and if a possible infringement is identified, the system triggers an alert. Tracking information and alerts are archived in the VMS server based at NEAFC and NAFO headquarters in London and Nova Scotia, and are sent automatically to surveillance teams operating in the fishing areas that are responsible for monitoring agreements. Based on information from the system, if a boat is found to be operating in a restricted area, the surveillance teams can deploy a vessel or an aircraft to investigate the infringement. By helping to focus surveillance activities in this way, the system permits far more effective monitoring of fishing activities in the area.

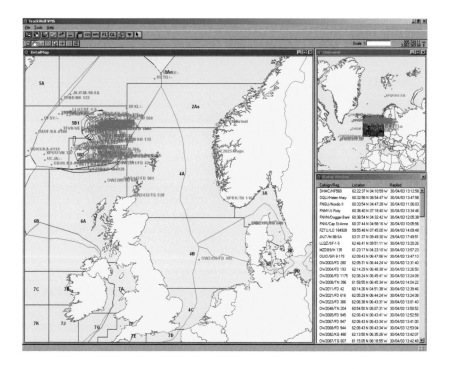

The Vessel Monitoring System interface enables current location of fishing vessels operating within convention areas to be displayed, as green dots, with an annotation giving the vessel call sign or registration number. The righthand text pane gives exact coordinate fixes and time of last transmission. For a given vessel it is possible to display its recent movements, current speed, and bearing. The system automatically issues alerts if a vessel strays from agreed exclusion zones.

Developed in the mid-1990s, the VMS had to meet a number of specific criteria. One of these was that the system had to be compatible with all communication systems used by the vessels operating in the area. Since these vessels came from all over the world, and used technology on board that ranged from basic to very advanced, this requirement was a significant challenge. It prompted a highly modular system design which permitted a very wide range of communication channels to be used, and allowed the easy development and integration of additional communication modules as new technology appeared. The system currently supports a very wide range of communications technology, including VHF/UHF radio systems, TETRA, Cellular (GSM and NMT), and satellite systems such as Inmarsat, EMSAT, Argos, Global Star, and Iridium.

The other critical requirement was security. The information being handled and transferred by the system is highly sensitive, both commercially and legally. Used by all participants of NAFO and NEAFC, the system needed to permit secure international access and data transmission. A Web-based viewing and data dissemination service was considered, but at that time, concerns about the security of data transfer across the Internet and the limited capabilities of Web-based mapping meant that the initial VMS was established as a traditional desktop application. It was based on Oracle and ArcView, and communication between server and remote sites used X.25 and X.400 network protocols.

THE DEVELOPMENT OF THE TRACSCAPE WEB SERVICE

As the mobile telecommunications market took off in the late 1990s, the modular nature of the VMS made it highly scalable and relatively easy to take key components from the system and establish them as Web-based services offered to commercial users.

Early tracking systems used for commercial fleet management (for example by trucking, delivery, bus, and taxi companies) were generally unique systems implemented separately by each company. This meant that each had to establish and maintain its own fleet administration control servers, display technology and (particularly in smaller urban areas) even dedicated communication networks and messaging systems. A Web service that could be offered as a standard solution by telecom providers and which handled all basic tracking and administration functionality offered a far more efficient solution.

Trackwell re-engineered the core tracking engine and communications modules of the VMS, giving them an open XML API so that they could be easily integrated with other third-party applications. A completely new Web-based administration and visualization module was written in Java, enabling all core monitoring functions—fleet administration, movement monitoring, alert notification—to be provided as a series of independent Web-based applications. With an XML API, integration with alternative network position finding systems (such as Nokia mPosition or Ericsson's MPS), establishing e-commerce payment gateways and access control modules that could be used by the host organization, was straightforward. In addition, the system could be easily integrated with a user's inventory or dispatch and distribution systems.

The tracking engine and communications modules of the VMS were re-engineered to form a flexible Web service. The communications and services layers can be expanded to add new services or communication protocols. The end user can access the system through the Web, through desktop applications, or through mobile devices such as PDAs and mobile phones.

TracScape Architecture

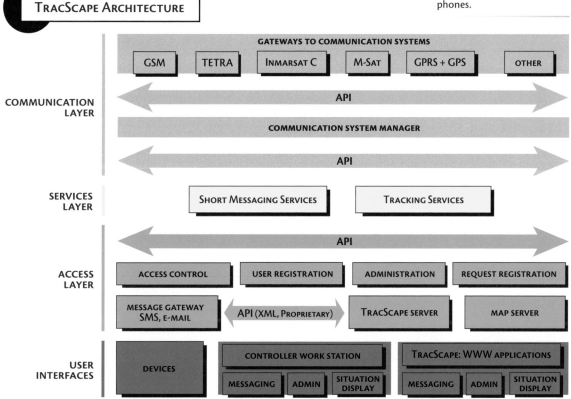

GATEWAYS TO COMMUNICATION SYSTEMS

| GSM | TETRA | Inmarsat C | M-Sat | GPRS + GPS | OTHER |

COMMUNICATION LAYER

← API →

COMMUNICATION SYSTEM MANAGER

← API →

SERVICES LAYER

Short Messaging Services | Tracking Services

← API →

ACCESS LAYER

Access Control | User Registration | Administration | Request Registration

Message Gateway SMS, E-mail | ← API (XML, Proprietary) → | TracScape Server | Map Server

USER INTERFACES

Devices | Controller Work Station | TracScape: WWW applications

Messaging | Admin | Situation Display | Messaging | Admin | Situation Display

The TracScape service enables both administrative activities—adding new vehicles, checking ownership, messaging—and tracking activities to be done through the Web. The service can be hosted by telecom providers or third-party dedicated tracking service providers or by customers themselves.

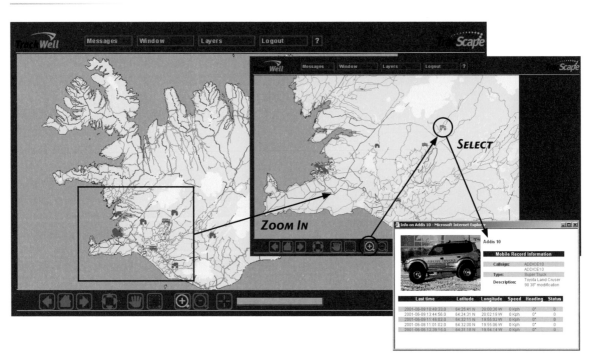

If an organization wanted, it could still take on the responsibility of hosting the services locally. However, one of the major benefits of TrackWell's system was that it opened the possibility for telecommunication providers to host the service to multiple users. End users, without having to concern themselves with implementation or maintenance of a 24/7 administration server or with display customization, could manage and track fleet movements with no more than a standard Internet browser.

The same hosted service can, for example, be used to track tourist adventure jeeps as they bounce across glaciers of central Iceland, the routes of taxis in distant Helsinki, or commercial freighters as they plow through the Atlantic waves. Each company has different needs and different communication requirements. For the tour company, the service provides the additional safety of knowing exactly where vehicles are in case of emergency and enables them to quickly plan and reorganize schedules in the (frequent) event of deteriorating weather. Communication for them means GPS navigation units carried on board the vehicles, and a TETRA-based communication module. For the taxi company, this kind of continuous monitoring is unnecessary; the system is needed only when drivers need urgent assistance and is linked to a panic button inside the cab that alerts the head office and police of the cab's current location through an SMS message. For the commercial freighter, the system enables the operating company and its clients to check the location of vessels at any time using an Inmarsat-C satellite communication to transmit current location to the receiver.

The telecom provider hosting the service can establish a customized environment for each client. Unique databases can be established, defining with the particular parameters relevant for the given business, the communication systems used. It also allows the look and feel of the user interface presented to the client to be customized. Behind the databases and custom interfaces are a series of services handling tracking, communication, database update, and map visualization operations. Though these services can take a wide range of parameters to match with the customized environment of each client, they are in fact standard services—the same service operating for all clients, simply handling different parameters.

TETRA

▸ TETRA, standing for TErrestrial Trunked RAdio, is a new, open digital radio standard designed for the professional mobile radio communications market (police forces, emergency services, transport companies, and so on). Unlike its counterpart GSM, used primarily in the consumer market, TETRA provides very high throughput of voice and data messages and the ability to manage radio spectrum to maximize efficiency and security. An open, widely adopted standard, it provides great flexibility in handsets and terminals used while providing reliable, cost-effective, and totally secure communications.

The end user has a Web-based interface both for fleet administration and for tracking fleet movements. Vehicles or vessels to be tracked can be uploaded into the database and alert zones, restrictions, and trigger rules added and amended. The tracking interface presents current location, route, speed, and direction of vehicles being monitored. Information can be accessed from any location that has an Internet connection, or from mobile units.

For the majority of users, Web services using HTTP protocol works satisfactorily. Demand for additional security from some clients, such as banks, has meant that the HTTPS protocol has also been supported, providing a secure Web-based environment for fleet management. In late 2002, an HTTPS Web services-based version of the fishing fleet VMS was undergoing initial trials at the Icelandic Coastguard. If these prove successful, it may not be too much longer before the original VMS developed for the Icelandic Coastguard and NEAFC is migrated to a full Web-services-based platform.

What is behind TracScape?

The architecture behind TracScape is modular and flexible so it can be established in a variety of configurations depending on its deployment. Typical configurations include a database server storing individual tracking databases, and dedicated map and Internet servers. These can be linked to central billing and mobile location services, and be configured to access administration systems used by the hosting organization, or to fleet management system.

XML is used as the protocol for communication between services and between TracScape and external applications. The system runs on a central Oracle database using ESRI's ArcIMS for map services. Interface modules are all developed in Java.

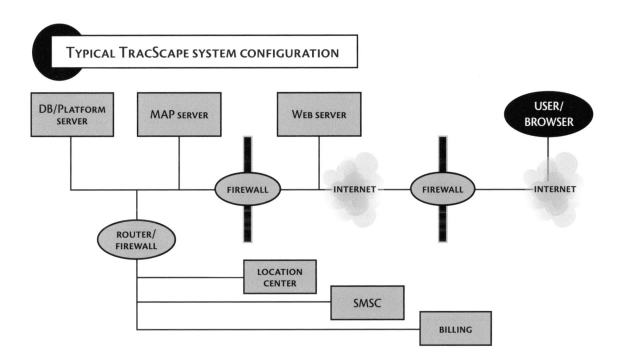

Typical TracScape system configuration

Dial M for Map

8

Providing location-based services (LBS) to the telecommunications industry has been a key catalyst behind the development of GIS Web services; it also happens to be one of the most challenging environments in which those services are deployed. On the one hand, demand—the need to provide location-relevant information to people on the move whenever and wherever required—has brought together diverse technologies, such as position finding, messaging, and spatial analysis and, as a result, has fostered the development of collaborative computing and Web services. On the other hand, the sheer size and pace of the mobile telecommunication boom meant that technology was changing almost monthly, with standards either nonexistent or perpetually in negotiation; developing services within such an environment inevitably leads to plenty of false starts and wasted effort. In addition, the limitations of both bandwidth and super-thin clients continue to pose real challenges for the transmission and presentation of map-based information.

Scandinavia is often seen as the incubator of the mobile telecom industry. Home to Nokia and Ericsson, some of the biggest names in the business, with traditions of industrial innovation and design that date to the nineteenth century, the area has been the incubator for many of the developments that have sustained the worldwide love affair with the mobile phone. Success breeds success, and the region has attracted a host of new cutting-edge companies and research labs engaged in exploring and expanding the possibilities of technology. Based in Oslo, Norway, Geodata AS has assembled a team that has been pioneering LBS provision to this sector. Working closely with telecom operators in Norway and elsewhere in Europe, in 2001 Geodata was one of the first companies in the world to offer third-party LBS GIS Web services for Wireless Application Protocol (WAP) based handsets, followed a year later with the release of a Short Messaging Service (SMS) GIS solution. This chapter examines the role of GIS Web services in the mobile telecom industry and looks at a selection of services that can be offered.

Geodata can be found on the Web at *www.geodata.no*.

To understand the role of Web Service in the LBS market one needs to have a look at the development of LBS as an industry. As telecommunications moved into the age of the mobile, "location"—specifically the ability to track where calls were being made—became essential for telecom operators. Although fixed-line telephone networks in which potential call

LOCATING MOBILE USERS

▷ There are a number of approaches to locating mobile users, providing differing degrees of sophistication and accuracy. They may be grouped into three basic types—data-based, network-based, and handset-based—depending on how the location fix is calculated. Variations within each of these types relate to whether or not signal speed, angle, or differences in time of arrival at different base stations are used.

Type	Method and strengths	Technology
Data-based	Location calculated with reference to a single base station. Fast and cheap but reliant on density of base station/cell-size distribution and as a result often provides limited accuracy.	Cell-ID, Time Advance, MAHO/NMR.
Network-based	Location calculated based on triangulation of three base stations. Fast and accurate, but increased overheads.	TOA, TDOA, AOA
Handset-based	Location calculated by handset-based GPS receiver or packet switching capabilities. Very accurate position fix, increasingly affordable.	GPS, A-GPS, E-OTD

volume at any point in the network can be relatively easily estimated and anticipated because the maximum number of lines connecting to any one switch is known and fixed, a mobile network is different. On a mobile network, phone users roam freely, so network demand is dynamic and very hard to gauge accurately. Locating where users were within the network and where phone calls were being initiated became essential for planning, in balancing capacity, and for maintaining levels of service delivery. Companies such as Nokia and Ericsson therefore developed technology to establish the location of mobile users based on their relation to an operator's network of receiving stations. One way of doing this was by allocating users to predefined base station cells. Another was by analyzing the speed and angle of a phone signal arriving at one or more stations.

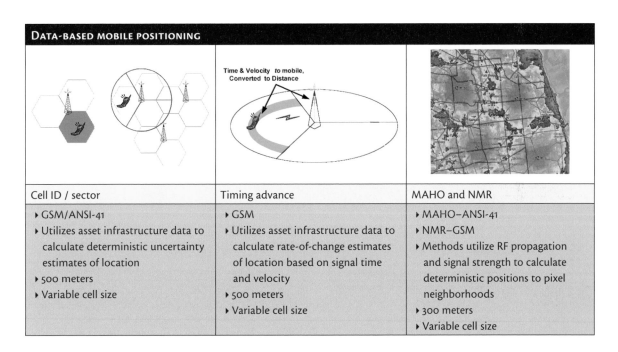

DATA-BASED MOBILE POSITIONING		
Cell ID / sector	Timing advance	MAHO and NMR
▸ GSM/ANSI-41 ▸ Utilizes asset infrastructure data to calculate deterministic uncertainty estimates of location ▸ 500 meters ▸ Variable cell size	▸ GSM ▸ Utilizes asset infrastructure data to calculate rate-of-change estimates of location based on signal time and velocity ▸ 500 meters ▸ Variable cell size	▸ MAHO–ANSI-41 ▸ NMR–GSM ▸ Methods utilize RF propagation and signal strength to calculate deterministic positions to pixel neighborhoods ▸ 300 meters ▸ Variable cell size

NETWORK-BASED MOBILE POSITIONING

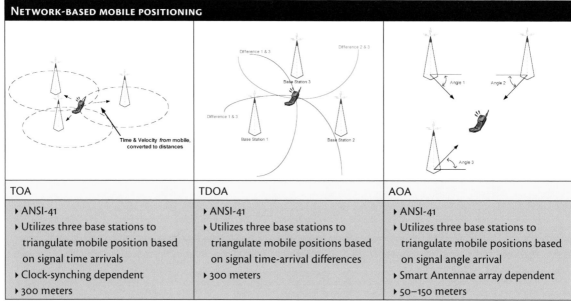

TOA	TDOA	AOA
‣ ANSI-41	‣ ANSI-41	‣ ANSI-41
‣ Utilizes three base stations to triangulate mobile position based on signal time arrivals	‣ Utilizes three base stations to triangulate mobile positions based on signal time-arrival differences	‣ Utilizes three base stations to triangulate mobile positions based on signal angle arrival
‣ Clock-synching dependent	‣ 300 meters	‣ Smart Antennae array dependent
‣ 300 meters		‣ 50–150 meters

HANDSET-BASED MOBILE POSITIONING

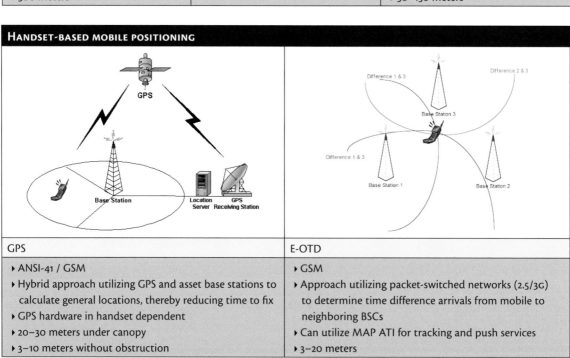

GPS	E-OTD
‣ ANSI-41 / GSM	‣ GSM
‣ Hybrid approach utilizing GPS and asset base stations to calculate general locations, thereby reducing time to fix	‣ Approach utilizing packet-switched networks (2.5/3G) to determine time difference arrivals from mobile to neighboring BSCs
‣ GPS hardware in handset dependent	‣ Can utilize MAP ATI for tracking and push services
‣ 20–30 meters under canopy	‣ 3–20 meters
‣ 3–10 meters without obstruction	

Though network load balancing prompted the initial development of this technology, it quickly became apparent that the information produced could be served back to the user as an entirely new and unique set of consumer services. A user with no more than a mobile phone or PDA could receive information and help that was specifically targeted based on their current location: if lost on an unfamiliar road, a map or driving instructions could be downloaded to a mobile phone; if a user wanted details of the nearest drugstore and how to get there, a text message to route the user and inform him of that week's sale could be created. If the user were injured and unable to move on a remote mountainside, a cell phone could provide the emergency services with his or her current location, greatly simplifying the search and rescue operation.

To perform any of these tasks, the location fix calculated by the mobile operator had to be passed to some form of GIS, or spatially enabled application, that could compare the fix against spatial data sets, such as address, point-of-interest, or route networks, and then implement spatial operations such as geocoding, proximity searches, and shortest-path routing.

Initially, in the late 1990s, telecom operators expected spatial application developers to establish a new interface that communicated directly with their proprietary position-fixing technology. However, for an LBS to be offered to the customer, the application would also need to interface to all the other administrative and financial systems, such as customer billing, security, and privacy, that made up the telecom's service. More often than not, that meant that the LBS application had to reside behind the telecom operator's firewall, located physically within their offices. With few standards governing data transfer and interface protocols, and with each telecom operator in effect having its own unique system, this was a complex and inefficient solution. Moreover, telecom operators would have to get involved in managing spatial data sets and GIS functionality. Third-party GIS application developers would have to develop LBS systems that were unique for each operator; since these were so tightly bound into the operator's other administrative and service applications, an upgrade, revision, or addition of any part, whether related to LBS or not, would inevitably require checking and modification in all other parts of the operation.

The steps taken to resolve this situation go hand-in-hand with the development of the concept of Web services architecture. Rather than integrating GIS tools directly into operators' systems, attention was focused on decoupling them, and on standardizing the format used for communicating among systems. LBS services began to be offered by third-party providers as Web services rather than being embedded in the telecom's own systems, supported by advances such as a flexible data definition syntax—eXtensible Markup Language (XML)—an increased awareness on the utility of the Web as a means of linking applications and data, and the development of standards governing message transmission—Simple Object Access Protocol (SOAP).

With these protocols in place, third-party spatial service providers can concentrate on developing new, innovative services within their chosen software technology secure in the knowledge that as long as their Application Program Interface (API) conforms to these standards they will be able to offer services to any number of telecom providers. Equally positive for telecom operators, this system means they no longer need to worry about maintaining spatial services themselves and can pick and choose from a range of services from any number of third-party companies without significant interfacing problems.

Geodata established its first Web service LBS in mid-2000 and has been continually expanding services since. Core spatial functionality is provided using a suite of GIS applications based around ESRI software. ArcSDE is used for data management and ArcIMS for map presentation and manipulation. Most spatial queries and operations are executed directly through the ArcSDE API, although ESRI's NetEngine™ is used for routing and proximity searches. Data used in the system may come from data sets maintained by Geodata or be pulled directly in from other Web-based data services.

Services can be provided directly to the mobile operators or, frequently, are provided to a dedicated "content provider" who may add to them or repackage them with other offerings before passing them on to the operator. What holds this very flexible environment together is conformance to Web standards—virtually all messages between parties use XML and SOAP; in those cases where SOAP is not supported, XML formatted messages are sent using an HTTP protocol.

This means that a single service hosted by Geodata—such as a proximity search—can be provided in multiple forms to different clients. To one client it can be provided as a complete service in itself. For example, an operator could pass position fixes to Geodata with a request to identify hotels in the neighborhood; the service would then process the request and pass back an SMS text message containing hotel addresses. For another operator, position fixes might be passed through a content provider who needs a basic route map to the nearest hotel and for the results to be returned for display on a WAP-enabled handset. In that case, the Geodata service would run the same proximity search, pass the output on to a routing engine to determine the nearest hotel and suitable route to it, and then pass a map back to the content provider. The provider could then add further detail, such as room rates or a description of the hotel that would then be passed back to the operator, and on to the user.

In one scenario, the telecom operator may receive requests to find the nearest hotel from both SMS- and WAP-based clients. Both requests can be handled by the same Geodata proximity service. In responding to the query, the Geodata service may access data or services from external sources such as point-of-interest or street databases. The response from the GIS Web service could also be integrated with third-party content providers—information on the selections may be pulled from tourist sites, for example. The complexity of these interfaces is greatly simplified through the adoption of standard XML and SOAP protocols.

LBS with Geodata for Operators and Content Providers

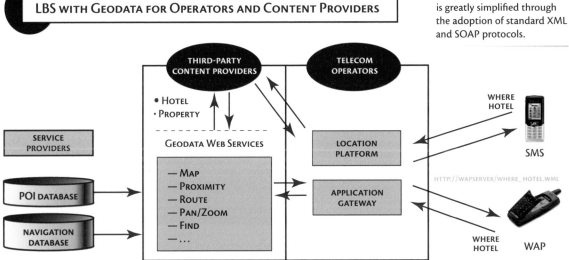

This has brought great flexibility to the implementation of LBS solutions for mobile operators, and has allowed companies such as Geodata to offer their services to multiple operators at any one time. In this way the Geodata Web service can be made available in many different European countries through multiple telecom operators and service providers.

SPATIAL SERVICES ON THE MOVE

Output from Geodata's services can now be presented to the user as maps on a variety of WAP, Multimedia Messaging (MMS) and 3G phones. On traditional handsets, SMS messages are used. Though WAP and MMS standardize how graphical data can be transmitted to mobile devices, map presentation on such terminals remains a major challenge, both in terms of map rendering (ensuring maps and annotation are legible) and in the wide variation in the size and resolution of the screens in mobile

SMS REQUESTS AND RESPONSES

▸ Typical examples of SMS requests and responses are shown below. The SMS keyword request is the request selected or keyed into their phone by the user. In order to provide some of these responses, it may be necessary to access services and data sets hosted by other third parties. Typical output provides examples of the kind of detailed response provided by these services despite the 160-character limit.

Service type	SMS keyword request	Data sets accessed	Typical output
Where am I?	Where	Current location Ward boundaries	You are in Skøyen (Oslo)
Find me the nearest *xxx*	Where *bank*	Current location Ward boundaries Point of interest	Near Skøyen (Oslo): XYZBank, Karenlystalle 7, ABC Bank Thomas Heftyes Plass . . .
Tell me how far it is to . . .	How far *Hemsedal*	Current location Ward boundaries Route network Traffic status service	You are in Sollihøgda and it is approx. 434 km to Hemsedal. Fastest route along E18, then Rv7.
Tell me the weather forecast	Weather	Current location Weather service	21/3 Hemsedal is sunny, 21C, SE5ms. 22/3 cloudy, pm rain, 13C

devices. Preparation of SMS messages is also complex; although the device specifications tend to be more standard, the return string length is limited to only 160 characters, so text messages must be carefully worded to be of any value.

Services provided all revolve around basic spatial functionality: map generation, proximity searches, routing, geocoding, and reverse-geocoding (taking a position fix and returning a textual address description). Although there are a host of different ways that these can be displayed to the user, Geodata's services address three basic user requests: *Tell me something about my current location; tell me how to get from my current location to another location;* and *tell me about other users near me.*

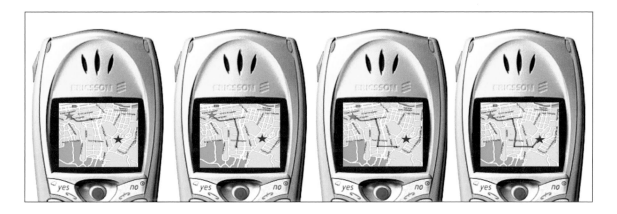

A typical example of a WAP-based route navigation service which identifies the walk or drive route between two points.

The first, providing information about current location, may be a simple task, but, if provided in sufficient detail, can be extremely useful—especially to a driver who has broken down on a remote road, and who learns that he is located on Route 344 and is 12 miles south of Hammerdal. As a variation, some carriers have also enabled the information returned to be copied to a given e-mail address, so the stranded motorist can forward his location directly to emergency services. This kind of local information service may also serve local weather forecasts or details of facilities such as banks, restaurants, or coffee shops within a given radius of the user's location. This has been established in some countries as a marketing tool for businesses. For example, a national

chain of gas stations may request a service that routes users to their nearest station, or modifies the content of a broadcast marketing message based on users' locations.

Routing services build on this basic operation. They can be used to provide walking or driving plans, distances, and estimated travel times from the user's location to the desired destination. They can also be tied in to data sets that maintain information on current road conditions, routing users on the move around congestion or road construction areas. Knowing which users are in the local area, or where a particular user is currently located can be used by parents keeping track of children, or in local chat or game services offered by content providers as a premium service.

For the end user, LBS has become just another set of services they have come to expect from their mobile operator. The ease and speed with which maps and directions appear on their handsets masks the great complexity involved in producing them—the multiple exchanges between different applications, written in different languages on different operating systems in different locations, the detail of the data sets accessed, the complexity of the business relationships and accounting systems that support.

It is the search for a route through this complexity that has accelerated the rapid evolution of Web services architecture, and its standards— a result quite appropriate to the task.

Mapping the News
Providing Web service mapping to the media industry

9

NEWS NEVER STOPS. The media, from the small local television station to the multinational media giant, depends on reliable, up-to-date information twenty-four hours a day, 365 days a year. News not only occurs at any time, it occurs anywhere—the most obscure street corner or the most remote stretch of coastline can suddenly become the focus of the entire world's gaze. To media organizations, accurate, up-to-date geographic information is essential to support and illustrate the stories they present to an increasingly well-traveled and graphically sophisticated readership. Operating on a global scale, without knowing where or when the next story will break, today's media organizations struggle constantly to find and maintain accurate and up-to-date geographic information.

Founded in 1848, The Associated Press (AP) is the world's oldest and largest newsgathering organization, providing content to more than fifteen thousand news outlets with a daily reach of 1 billion people around the world. Its multimedia services are distributed by satellite and the Internet to more than 120 nations. AP provides image and video libraries, editorial copy from more than 240 bureaus around the world, extensive archives, and, since late 2000, MapShop—a simple, easy-to-use Web service that gives subscribers access to up-to-date mapping for any location in the world around the clock.

LOCATION: AN ESSENTIAL PART OF NEWS CONTENT

How to explain the logistics of food aid distribution across the deserts, rivers, or mountains of an unfamiliar country? How to illustrate the tangled terrain of a distant country where international peacekeeping troops are now maneuvering? How to reveal a new political landscape unfolding minute by minute on election night? How to highlight the extent and consequences of global warming, such as the destruction of primary forests, or sea-level rise? How to inform the public of the route of a royal procession or a marathon race? How to present tomorrow's weather?

Words are not always enough. In the highly competitive print media industry the inclusion of eye-catching, high-quality, colorful graphics to support the written word is essential. For stories like those outlined above, this means maps.

The relationship between the media and maps has been a long one. Newspapers such as *The Times* of London and the *New York Illustrated News* have, since the nineteenth century, been a common source of publicly available maps and atlases and *The Times Atlas of the World*, for example, remains one of the most respected cartographic and gazetteer publications in the world. However, the resources required to sponsor or collect such information are, in the twenty-first century, well beyond all but the most specialized or well-funded publications. The media industry has been increasingly turning to third-party data sets, graphics, and mapping packages to satisfy demands for detailed and flexible cartographic presentations.

EARLY MAP ILLUSTRATION

▷ *The Times* routinely published maps illustrating its stories from the mid-nineteenth century onwards. The first *Times Atlas* was printed in 1895 as a set of weekly installments spread over fifteen weeks. Each installment cost 1 shilling (about $4.60 U.S. in today's terms). The complete atlas consisted of 117 pages of maps and a gazetteer of 125,000 place names "brought up to date and specially prepared for *The Times*." (By comparison, the current edition contains more than double the number of map pages and well over 200,000 place names.) It was in fact largely a reprint of an earlier atlas issued by Cassell & Co., but was a great success, capitalizing on *The Times'* extensive distribution and marketing network. Later editions were specifically commissioned, for example the 1922 *Times Survey Atlas of the World*, which was produced in conjunction with John Bartholomew & Co. to provide readers with a comprehensive atlas updated to reflect the major boundary changes after World War I. In the United States, interest in the unfolding Civil War meant that leading publications of the day (such as *The New York Illustrated News*, *Harper's Weekly*, and *The New York Daily Tribune*) employed large numbers of cartographers and competed with each other to produce ever more striking, up-to-date maps of troop movements and recent engagements. Rand McNally came into being in 1864 when the *Chicago Tribune's* job print shop was purchased by William Rand and Andrew McNally who then built the business serving both the media and individual buyers with maps and atlases covering the westward expansion of the railroads and commerce, and international events such as the American Spanish War.

The reasons for this are several. Few media organizations can muster full GIS teams, or have the available resources to subscribe to the wide range of digital data sets that may need to be referenced for map preparation. The majority of map work is undertaken by general graphics departments with relatively little cartographic training and, usually, very limited time; graphics departments quite frequently have no more than an hour to respond to a particular request. As a result, there has been a tendency for newspaper maps to be rushed through, based on standard data sets with only minor additions or modifications—often sketched in by eye—for any particular story. Innovative use of map techniques—projection, orientation, annotation, shading, or 3-D effects—to illustrate a story has been heavily restricted and generally only used for stories that could be forecast several months in advance, for example, coverage of a regular sporting event or an election.

As GIS Web services were being developed, The Associated Press realized there was potential for providing its members with a Web-based mapping service that would not only give graphics departments a tool to put together accurate, sophisticated maps quickly, but would also provide access to a vast collection of online up-to-date data.

An example of a map graphic package that AP sends to member newspapers. It includes not only the high-quality map itself that will be reproduced in the local newspaper, but also essential details the newspaper's editors will need, such as the source of the map data, the name of the accompanying news story that it illustrates, and supporting information for informing newspaper readers of the map content.

Associated Press
graphics

<AP> **CHILDREN POVERTY 102902:** Map shows results of a survey on the percentage of school-aged children living in poverty; 2c x 4 inches; 96 mm x 102 mm; with BC-Children-Poverty; ETA 7:30 p.m.</AP>

South and Midwest highest in poverty

The South and Midwest have the highest concentration of children living in poverty, according to Census Bureau estimates released Tuesday.

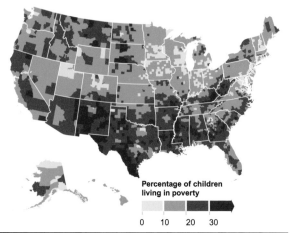

Percentage of children living in poverty

0   10   20   30

SOURCE: U.S. Census Bureau   AP

Editor's note: It is mandatory to include all sources that accompany this graphic when repurposing or editing it for publication.

MapShop services

MapShop allows reporters, editors, graphic artists, producers, and anyone else with an interest in disseminating maps to log onto the Web and access terabytes of geographic information covering every country in the world.

The mapping interface provides a range of spatial search and map design tools. These include simple feature-, place name-, or address-based searches, as well as the customary pan, zoom, and scale map navigation tools. Journalists can select which feature layers to display and how they are to be presented. Symbol and font selection functions provided within MapShop are some of the most sophisticated available in Web-based mapping tools. Users can select from a wide range of standard line, shade, and marker symbols, or create new ones online. In addition, they can load their own font symbols, which can be seamlessly integrated within MapShop. Style libraries are provided to facilitate rapid creation of quality map products, and user preferences can be saved to standard templates so that maps can be quickly formatted to a newspaper's individual style.

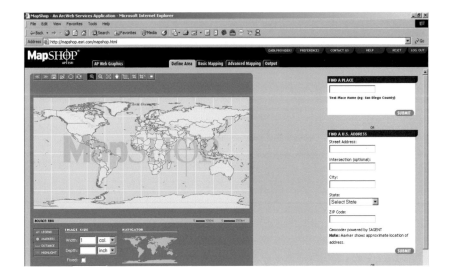

The interface that MapShop subscribers use to get started is presented in easy-to-understand fashion, with users given several different ways to narrow their search for geographic information about a particular area in the world.

MapShop automates the placement of annotation, the generation of latitude and longitude lines, and the selection of appropriate layers for the scale at which the map is being viewed. In addition, it permits the projection of the map to be modified on the fly. Annotation, both text and symbols, can be added to the maps, as can legends, labels, and scale bars.

The entire map design process can therefore be undertaken within MapShop in a matter of minutes rather than hours. The finished map can then be downloaded in a range of formats suitable for direct integration with desktop publishing or graphics editing software. If necessary, the map data can be downloaded to local GIS software (ArcView or ArcInfo) where it can be developed further and local data added.

On the left of the interface, the MapShop user has narrowed a search to the New York/New Jersey/ Connecticut metropolitan area. On the right, she can choose from several base map layers, and then, from the box below, which thematic layers available for the area that would best illustrate the information that needs to be conveyed.

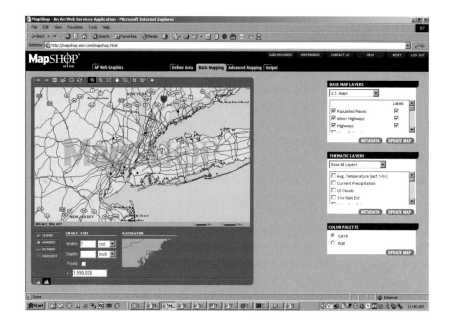

CONTENT AND SERVICE DELIVERY

Behind MapShop sits a wealth of data brought together and presented through an ArcIMS mapping interface.

Some of these data sets are freely available, but many of them normally require payment. Subscribers to MapShop enjoy full access to all data sets. Due to economies of scale, this vast resource can be offered to subscribers at a fraction of the cost that would be incurred if individual subscriptions were taken out with each provider.

Perhaps a more significant saving, however, stems from the speed and the ability of MapShop to automatically integrate these separate data sets into a single seamless database, without the need for an a deadline-pressured graphic artist or editor to be burdened with complex additional tasks such as registering images or converting projections. For an overburdened graphics department, the ability to create simple, accurate, visually appealing maps within minutes is an enormous benefit.

Data management and integration is handled by ESRI's ArcWeb Services platform, acting as the primary data broker connecting and integrating the services offered by different data suppliers. The MapShop interface is developed in Java and ArcIMS.

A MAPPING SOLUTION FOR A MAJOR NEWSPAPER

One of the earliest organizations to start using MapShop was *USA Today*. With a circulation of 2.3 million (available in more than sixty countries), *USA Today* is the largest selling daily newspaper in the United States. Striking, clear graphics in both color and black and white are one of the hallmarks of its style.

USA Today subscribes to a number of AP media services and was quick to realize the potential of MapShop. It became a beta tester in 2000. The system was really put to the test in late in 2001, when the graphics team was faced with particularly urgent, competing demands. In any one day the small graphics team produces dozens of illustrative maps and graphics and runs at full stretch most of the time. In addition to this normal load there was an urgent request for two new sets of topographic maps. One was to provide maps on which to chart the unfolding military campaign in Afghanistan and illustrate the rugged terrain in which U.S. armed forces were beginning to operate around Kabul and in the mountains to the north and east. The other was a series of venue maps to illustrate feature stories from the upcoming Winter Olympic Games in Salt Lake City, Utah. Both could be illustrated with simple contour maps, but with such important stories *USA Today*'s editors were looking for something much more striking.

The answer was relief shading—shading terrain illuminated from an imaginary light source to highlight the structure of its slopes and valleys. Relief shading, however, is an extremely time-consuming exercise, particularly if it were to be done for multimap series of the complex topography in Utah and Afghanistan. Using traditional techniques this would require weeks of work, time which was simply not available. With MapShop it became a feasible option.

Salt Lake City is situated among the peaks of the Rocky Mountains, making a graphic representation of its geography by MapShop illustrators a challenge.

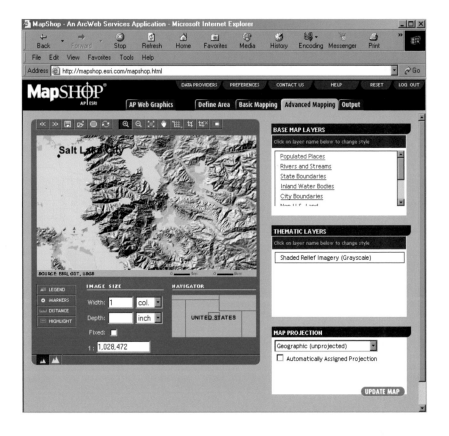

RELIEF SHADING

▷ Relief shading, one of the most intuitive ways of drawing and visualizing three-dimensional surfaces, has a long history. The seventeeth-century painter, mathematician, and cartographer Hans Conrad Gyger (1599–1674) advanced the technique and spent thirty-eight years completing his 1 : 32,000 relief-shaded map of the environs of Zurich. Since then techniques have been refined by a number of leading cartographers around the world including recently, the Swiss cartographer Eduard Imhof (1895–1986) and the Austrian Heinrich Berann (1915–1999). Computerized techniques for relief shading continue to evolve. For more information see *www.reliefshading.com* and *www.nacis.org/cp/cp28/ resources.html.*

Within a couple of hours both sets of relief-shaded maps were produced—to a level of detail and accuracy that would have been unimaginable using the resources previously available.

Making use of up-to-date terrain maps from Worldsat International and Chalk Butte, Inc., the paper's graphics team quickly produced a relief-shaded base map. MapShop also provided an automatically generated legend, scale bar, and correctly positioned latitude and longitude lines. They used other MapShop resources to add major place names, water bodies, and key roads. This was then transferred to *USA Today*'s existing graphics systems, where staffers added some of their own data and information collected specifically for the stories, before passing it to the editorial team for inclusion on the page layout. The richness of the MapShop data sets—many of which span the globe—and the flexibility and ease with which these could be integrated with their existing graphics and publishing systems were a revelation.

MAPSHOP: CONNECTING DATA PROVIDERS WITH THE MEDIA INDUSTRY

MapShop is an excellent demonstration of how Web services can be used to bring customers and providers together.

The majority of the data services offered through MapShop was available prior to its launch, as were the GIS and map-building capabilities that it provides. However, each was provided as a discrete service that prospective users had to locate, negotiate, and then integrate, both with each other and with their existing networks and software. MapShop does all of that for them, providing a single, highly accessible, targeted application.

For the data providers, MapShop offers a useful way of presenting their products to a specialized market, which may otherwise be difficult to penetrate. For the users, economies of scale provide significant cost savings in subscription charges for individual data sets; in addition, data maintenance, updates, and distribution are taken care of, and data is provided in a simple, easy-to-use mapping tool.

Based entirely on Web linkages and on the loosely bound, ever-expanding infrastructure of the ArcWeb Services platform, MapShop is working to bring up-to-date spatial data to the media industry, and into our daily news.

Building Skills with Web-based Spatial Editing

10

GIS HAS GREAT POTENTIAL to help us understand the natural world, and to monitor and manage the way we interact with it. It is being used by international organizations, governments, conservation groups, and academics around the world to research some of the most pressing environmental problems facing us, including deforestation, desertification, and global warming. Key to the success of these efforts, however, are people trained in the basics of spatial analysis—people with an awareness of the potential (and limitations) of geographic data sets, who are capable of interpreting results, and of implementing and sustaining data collection and monitoring programs. As young a science as GIS is, this kind of knowledge is still often hard to come by, and must be nurtured through education programs and training of scientists and technicians from a diverse range of fields.

This chapter examines the success of one university in China where Web services are helping to expand GIS education, and to introduce a new group of scientists and engineers to the potential of GIS technology.

Water for a population of 1.5 billion is one of China's most urgent environmental concerns. The need for water, and for effective water management, is immense. Seventy percent of grain is grown on land that must be irrigated; continuing industrial and urban growth create a never-ending demand for reliable water supplies; drought and flood frequently plague large swathes of the country. Water issues reach beyond China as well;

The North China Institute of Water Conservation and Hydroelectric Power, based in Zhengzhou, Henan Province, is one of China's leading technical institutes for water management, conservancy, and engineering. Its current campus was build in 1990 and is home to more than nine thousand undergraduate and postgraduate students.

desertification, windblown dust, and siltation of key continental drainage channels affect the whole Asian region and beyond. Given the magnitude of these problems, policies for efficient, sustainable water management are being sought by both national and international organizations. One of China's leading technical institutes focusing on water management and conservancy is the North China Institute of Water Conservation and Hydroelectric Power.

The Institute, founded in 1951 under the auspices of the Ministry of Water Resources, was originally called the Beijing Water Conservation School and was housed in a single building. It has expanded greatly since, and moved to its current location in Zhengzhou in Henan Province on the banks of the Yellow River in 1990. More than nine thousand undergraduate and postgraduate students, as well as some of the leading Chinese researchers and professors in the field work and study at the Institute. Still working closely with the Ministries for Water Resources and Construction, the Institute trains many of the staff that go on to fill senior management and research positions within the Ministries' offices throughout the country.

The institute's educational curriculum covers a wide range of technical subjects including water management, conservation, and irrigation, as well as hydrology and related hydraulic, geotechnical, geological, and civil engineering disciplines. In 1999 it took advantage of ESRI support programs for educational institutions and invested in a suite of ESRI products, including ArcInfo, ArcView, ArcSDE, ArcIMS, ArcPad®, and MapObjects. It installed the technology in the Geotechnical Engineering Department, and the department became the focus of GIS teaching and learning at the institute.

Recognizing that GIS was relevant to students' professional development across many of its disciplines, the department began considering ways to allow as many students as possible to have access and instruction on its GIS facilities. With only twenty-six copies of ArcView, and single copies of the others, there was never enough software so that every student could be given real hands-on experience. Even with generous educational discounts on software and maintenance, expanding the number of users' licenses significantly was not a viable financial option for the institute.

As a partial solution, the department began experimenting with the creation of ArcIMS intranet applications for training purposes. These let large numbers of students get an idea of some of the technical data sets available, and allowed them to do simple spatial queries. However, such intranet solutions could not give students experience of detailed modeling or complex spatial analyses. More importantly for students who, after graduation, were likely to be posted out to remote field stations with responsibility for large data collection and monitoring projects, it could not give them hands-on experience of data set development and maintenance.

Despite these limitations, intranet-based training applications proved extremely popular and helped to introduce GIS to a far wider group of students than had previously been feasible. Based on this success, the department started to explore the possibility of expanding its applications to

Tig Centre can be found on the Web at *www.tigcentre.com*.

Dr. Tong Jiang

The WebEdit Web service enables data in shapefile format or stored in ArcSDE to be edited across the Web. At the Institute, the service is run on the intranet and helps expose students from many different departments to elementary principles of GIS editing and data management.

cover more advanced GIS functionality through the use of Web services that can give intranet access to advanced data maintenance, modeling, and analytical tools. The first element of this initiative was the introduction in late 2002 of TIG Centre's Web-based WebEdit (WE v2.0) tool kit. TIG Centre is a GIS development firm based in the United Kingdom and Hong Kong that specializes in research and development of Web-based GIS solutions.

WebEdit was added to the intranet teaching materials to develop students' awareness of issues relating to update and maintenance of spatial data sets. Through lab or fieldwork exercises, it is relatively easy to train students in basic data collection—land survey, geological or soils mapping, aerial photo interpretation, taking field measurements, designing sampling programs, and related applications. Either through use of hard-copy maps and statistical tables or through the intranet ArcIMS training applications, students could also be introduced to the final data sets produced or maintained by these data collection techniques. What was missing, however, was practical instruction in the assembly and selection of data collected in the field to update existing digital data sets or to produce new ones. Students can be lectured about the importance of updating data sets with compatible data, whether in terms of scale, projection, collection technique, or accuracy; on how accurate metadata saves time and expense both for users and data managers; and on how data entry and storage methods can affect the kinds of analyses that may be done in the future. But nothing beats practical experience as a means of highlighting the real import and the ramifications of these practices. That experience can be gained through hands-on work with GIS packages such as ArcView and ArcInfo, but the limited number of licenses held at the institute means that access to these packages can only be given to a relatively small number of students. WebEdit provides a way of introducing these concepts to a far wider student audience.

The WebEdit service provides a basic spatial data editing and administration tool kit. It works alongside ArcIMS to enable users to add and edit data stored in ESRI shapefile format or in any data set that can be accessed through ArcSDE (including spatial and textual data managed directly in ArcSDE, as well as data managed in third-party software such as Oracle Spatial). It enables data features such as lines, points, and polygons to be added or edited in a standard Internet browser and permits administrators to manage and review updates being made by multiple users.

Integrated with existing ArcIMS teaching applications, WebEdit has been deployed in a variety of ways.

Rather than simply using existing data sets presented through ArcIMS, students can now build their own using a complete suite of editing functions, including add, move, copy, delete, and rotate operations. They may, for example, be tasked to develop a data set of a complex river system. This would require them to make decisions on, for example, where in the network to start digitizing both river banks rather than a single line representing the river channel, how mid-stream islands or braided channels are captured, or how seasonal variation in river course or width is recorded. In building data for models, rather than simply talking about the importance and implications of network connectivity, students are required to build their own networks and can learn through experience that, for example, links must connect at nodes and directionality is important, and just how much effort they will have to expend to maintain such data.

When this kind of instruction is combined with courses on field data capture using traditional survey techniques or GPS, students can learn about a continuum of GIS work—from data gathering in the field to bringing their data into a digital environment; this enables them to work through issues involved in projection of surveyed data and integration with existing data sets. A COGO interface enables the capture of data from standard surveying techniques.

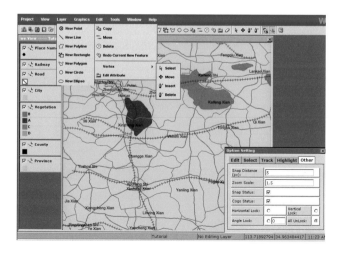

The WebEdit service enables users to add, move, edit, and delete spatial features, and provides basic edit controls such as feature snapping, record locking, and roll back. The service can be hosted in a number of different languages.

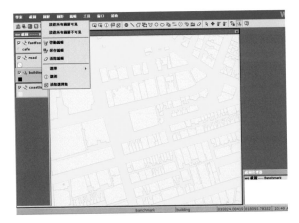

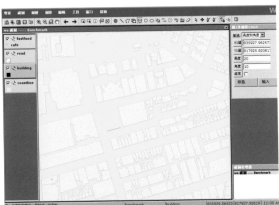

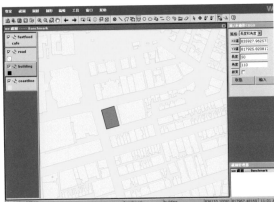

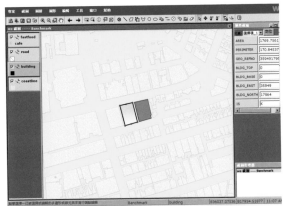

In this example, a student is updating a building data set: first, a building is added by specifying angles and distances, input in the COGO window to the right. This is then saved back to the server, and the cursor and vertex edit function is used to split the neighboring building into two freestanding structures. Textual attributes for each structure are being updated in the last image.

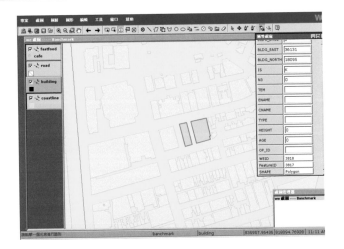

The WebEdit administration service controls the edit environment and acts as an interface between the editing user and the base data sets being updated. It can be used to illustrate approaches to concurrent or transaction editing, record locking, and resolution of conflicting updates. For example, an entire class can be assigned to work on data collection and update of a single data set, but one group may be designated as administrators, with responsibility to define the update strategies and the work program, and to monitor the progress of their classmates. The administration service can be used to control which data sets or fields users can access and update, and how updates are recorded and logged to the central database. Working through the transaction history, the class can review the update strategy, work through rollback, resolve conflicting updates, and assess the suitability of various update strategies.

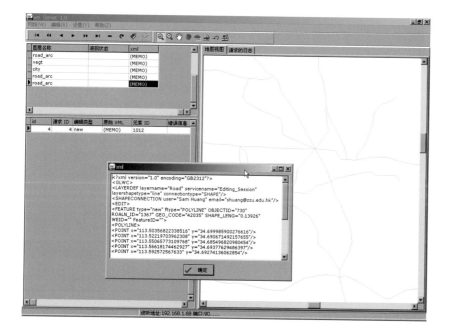

The WebEdit administration service enables edits to be reviewed and, if necessary, rolled back. All edits saved back to the central server are recorded and the related XML message can be examined.

Having to develop data sets rather than simply use existing ones gives students a far better understanding of how digital data represents features on the ground, and what is involved in data capture. Although WebEdit can be used only as an introduction to GIS and for data maintenance, it is extremely valuable in reinforcing principles of data format, update, and maintenance to a very large number of students. Allowing students to actually develop their own data sets is a far more effective way for them to explore these issues than using existing predefined data.

Through exposure to the combined ArcIMS WebEdit training tools, a growing number of undergraduate and postgraduate students have begun to explore and use the Geotechnical department's full GIS suite and to start incorporating GIS tools and data in their projects.

The WebEdit service is developed in DHTML and JavaScript™. The administration service runs alongside ArcIMS and handles data extraction from ArcSDE and shapefile data sources, transaction management, edit tracking, and conflict resolution. Edits and additions are recorded in Vector Markup Language (VML). The client service provides typical mapping tools (layer control, symbology, and map display) as well as editing tools. Communication throughout the system is based on XML and ArcXML.

The Institute runs WebEdit on its departmental intranet. Data is held in a number of data servers in the Geotechnical Engineering Department either as directories of shapefiles, or stored within a central SQL Server database and accessed through ArcSDE. The ArcSDE database runs on a Pentium 4 Xeon machine with a 20 GB hard disk. ArcIMS Image services and WebEdit Server are run on two separate machines running Microsoft Windows® 2000 and IIS 5.0. Client machines need only be on the university's intranet and to run Microsoft Internet Explorer 6 or higher. The data used comes from the institute's own projects and national data sets from the National Fundamental Geographic Information System (NFGIS), the State Bureau of Survey and Mapping (SBSM), the Chinese Seismology Bureau (CSB), and the Yellow River Conservancy Commission (YRCC).

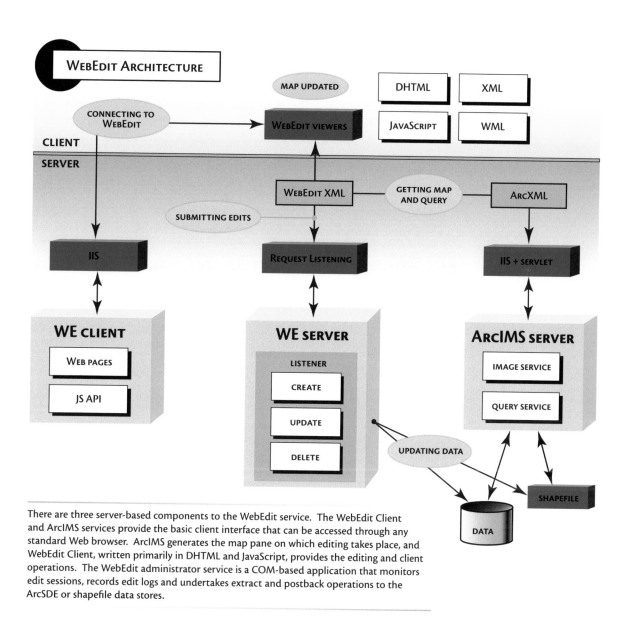

There are three server-based components to the WebEdit service. The WebEdit Client and ArcIMS services provide the basic client interface that can be accessed through any standard Web browser. ArcIMS generates the map pane on which editing takes place, and WebEdit Client, written primarily in DHTML and JavaScript, provides the editing and client operations. The WebEdit administrator service is a COM-based application that monitors edit sessions, records edit logs and undertakes extract and postback operations to the ArcSDE or shapefile data stores.

Encouraged by the success of the editing Web service, the Institute is currently exploring the feasibility of establishing some of its modeling and analytical tools as Web services which may be combined with the WebEdit ArcIMS teaching materials to provide a comprehensive introduction to GIS. The courses established so far have demonstrated the ability of Web services as a means to enable large numbers of students to gain hands-on experience with GIS tools. They have introduced a far greater number of students to basic GIS concepts than was ever possible before. As this program is expanded it will help to ensure that those at the forefront of China's efforts to conserve and manage water in a sustainable, responsible manner, whether engineers, surveyors, or hydrologists, will have a firm understanding of spatial data, GIS, and its potential to help their efforts.

Fueling the Oil and Gas Industry with Data

11

"CONTENT" WAS THE TRENDY, OVERWORKED BUZZWORD coined during the late 1990s dot-com boom to distinguish the unique characteristics of one new Web site from another, or to explain why one dot-com failed and another prospered. In many cases, content—synonymous with "valuable information that clients are willing to pay for"—was never there and never materialized. Thus, the boom became a bust.

The oil and gas industry, light years from the vaporware world of dot-com start-ups, has always understood the real value of content. Oil service and exploration companies frequently operate in some of the most inhospitable, remote parts of the world, where basic information on the shape of terrain or seafloor—let alone what lies beneath—is difficult to obtain. With huge upfront investments required before a drop of oil or gas starts flowing, companies in this industry know the importance of accurate, reliable data. Even when production starts, a successful operation relies on information about the complex logistics of downstream refinement and distribution to market. Content, whether it's exploration records from previous surveys, up-to-the-minute market prices in the world's capitals, or details of the latest export tariffs imposed at a foreign port—and the ability to effectively amass, reconcile, analyze, and interpret data—is all critical to success in this fiercely competitive business environment.

The IHS Energy Web site provides access to a full range of industry statistics, news, data sets, and reports, as well as providing a gateway to dedicated application, consultancy, and support services.

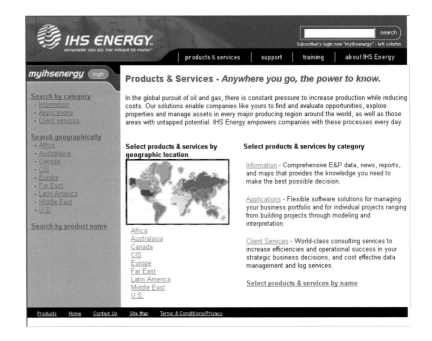

IHS Energy can be found on the Web at *www.ihsenergy.com*.

Collecting, consolidating, and serving data to the oil and gas industry has become a major business in itself. One of the biggest players is IHS Energy, which provides data, consultancy, research, and analytical support to the oil and gas industry throughout the world. Recognizing the potential of the Internet as a means of supporting its clients, IHS Energy has been developing its Internet- and Web-services-based information services since 1997.

IHS Energy's information resources are vast. Assembled through acquisition and through a global network of relationships with industry partners and contacts, they bring together data on petroleum exploration and production that covers the globe and dates back more than fifty years.

All major oil producing areas of the world are covered. IHS Energy gathers information on current exploration activity and maintains historical databases. Data covered includes wells, fields, production, reserves, concessions, geophysical surveys, partnerships, and basin-scale geological interpretations. Up-to-the-minute reports on worldwide activity are gathered through a network of contacts and consultants stationed in oil-producing areas across the world. Details on locations, capacities, and utilization of the world's ports, tankers, and pipelines, the regulatory structures of different countries, and fluctuations of demand and price are also recorded, as are forecasts and details of new developments and plans that will impact future production.

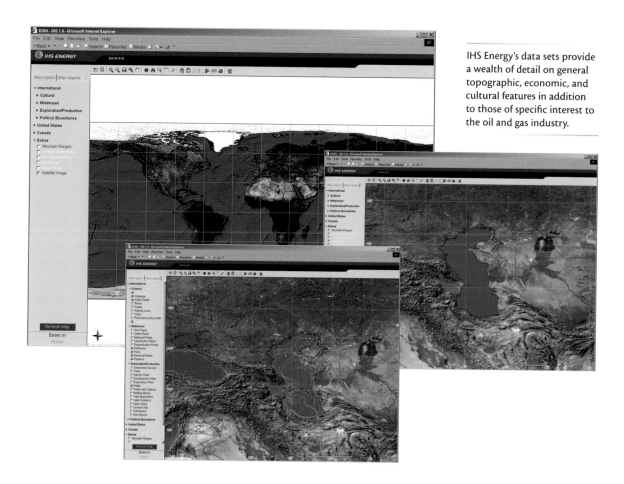

IHS Energy's data sets provide a wealth of detail on general topographic, economic, and cultural features in addition to those of specific interest to the oil and gas industry.

As can be imagined, the work to validate and collate these data sets into a coherent, reliable resource is significant. To accomplish it, IHS Energy relies on GIS as a means of assembling and managing diverse data sets. The company has spatially encoded much of its data, holding it in a series of very large Oracle databases managed through ArcSDE. The kind of effort involved in collecting and maintaining these data sets can, however, only be justified if the final product can be presented to the client in a readily accessible and meaningful manner.

In the past the only way to distribute these databases to clients around the world was on tape or CD–ROM. Data was distributed with a custom desktop application packages running on ArcView, ArcGIS, or other third-party software that permitted data to be queried, viewed, and extracted. These custom applications could provide powerful tools for quick retrieval and analysis of all major data sets, as well as for data extraction to a range of standard formats. Those wishing to develop their own analytical software or integrate the data sets with existing applications could do so by accessing data directly through Oracle and ArcSDE's open API.

The data sets provided by IHS Energy are, however, very large (totaling in excess of 100 GB). Distributing these on CD–ROM not only required users to invest in hardware and software to load these data sets locally in their offices, it also committed them to ongoing investments in staff training and in expenditures of time for tuning and active data management. Routine updates (normally monthly or quarterly) could total as much as 10 GB. In addition, integrating these data sets with other in-house applications, or making them available across the corporate intranet, while feasible, still required the client to undertake fairly major application development. As a result, access to these databases generally remained restricted to a small number of desktops. For IHS Energy this method required administrative overheads as well—in routinely extracting data from its databases, in production and distribution of CD–ROMs, and in support for clients managing their data locally. In addition, monthly or quarterly updates could not reveal the full richness of the data sets, many of which were updated on a daily basis. Most frustrating was the knowledge that the data was already managed in well-tuned databases in IHS Energy. If clients could directly access and work with these, they would not only have access to data as soon as it was validated and added to the master data sets, it would relieve data administration overheads both for the clients and IHS Energy.

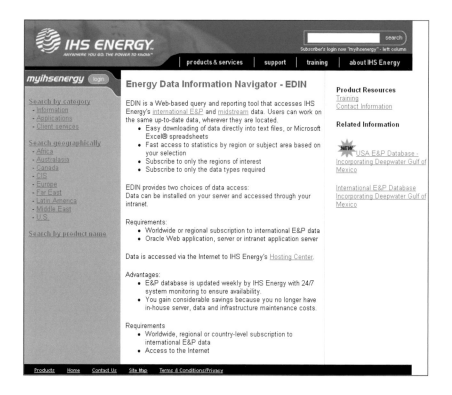

Internet delivery and Web services offered the potential of an alternative solution. IHS Energy watched the evolution of the technology carefully, acutely aware of concerns relating to performance and security that would need to be addressed before launching their data and services on the Web.

Their first Web-based product, Energy Data Information Navigator (EDIN), was released in 1997. This permitted Web-based access to two of IHS Energy's key databases, the International Exploration and Production Database—which provided prospecting results, well and concession locations, and production statistics. The other was the MidStream Database, which detailed worldwide transportation and refinery infrastructure. EDIN could be used to access tabular data hosted directly by IHS Energy through the Internet, or to facilitate intranet access across corporate networks for situations where IHS Energy data continued to be hosted locally by clients. A desktop GIS application (based on ArcView 3.x) was released

which enabled intranet-based EDIN users to access the geographic data stored in IHS Energy data sets. What was lacking was a way to provide direct access to the graphic GIS-based data across the Internet—the complexity of the data sets stored and the way in which they needed to be presented and analyzed was beyond Internet mapping at the time.

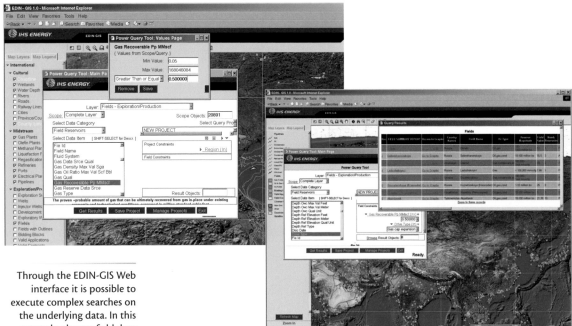

Through the EDIN-GIS Web interface it is possible to execute complex searches on the underlying data. In this example, the gas field data set can be filtered based on quantity of recoverable gas supplied by each field. Based on the search, summary tables and reports are presented which can be interrogated further to show spatial distribution or related graphs and reports. Multiple queries may be executed and the results and query parameters saved.

In 2002, the EDIN Web service was expanded to permit GIS mapping functionality to be accessed over the Internet.

EDIN-GIS provides a Web map service that can be used either to allow distributed intranet-based access to existing IHS Energy spatial data sets (if clients wish to continue to receive CD–ROM updates and maintain their own local copy of these data sets), or more significantly, to allow direct Internet access from anywhere in the world to data that is actually hosted by IHS Energy. The latter entirely removes the need to maintain data at the client site, greatly reduces dissemination, maintenance, and update overheads, and enables clients to access data as soon as it is updated in IHS Energy's central databases. The other advantage is that via the Web, EDIN-GIS can be accessed from anywhere with a reasonable Internet connection (56ᴋ line or faster) and is no longer tied to a specific number of desktops.

Users can, for example, log onto EDIN-GIS and explore the geology of a play (oil reservoir) around the Caspian Sea—first by checking the interpreted geophysics and geological maps from a variety of sources, then moving in to a particular area of interest, and retrieving detailed basin,

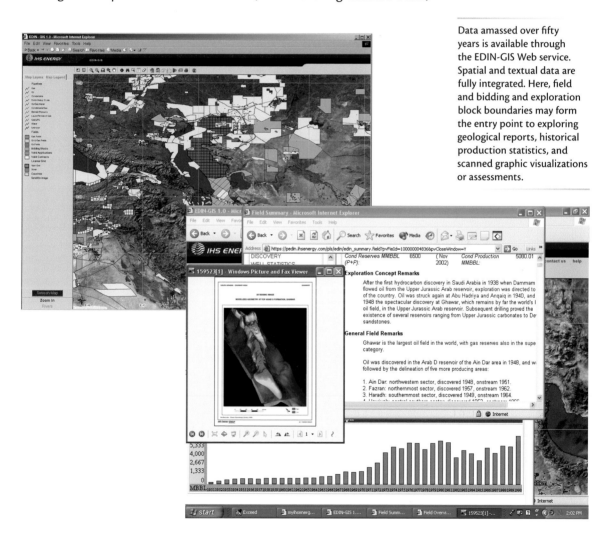

Data amassed over fifty years is available through the EDIN-GIS Web service. Spatial and textual data are fully integrated. Here, field and bidding and exploration block boundaries may form the entry point to exploring geological reports, historical production statistics, and scanned graphic visualizations or assessments.

well, and field reports. Data is presented in a variety of ways—thematic maps, tables, and reports and graphics. Such data will necessarily be supplemented by further on-site exploration, but ready access to data sets accumulated over fifty years provides an invaluable resource for high-level prospecting and validation of results.

The commercial or production potential of an area can be explored based on current or historical production statistics from wells operating nearby, as well as availability, ownership, utilization, and fee structures of nearby land, ports, transportation, and pipeline facilities. Background cultural information such as natural and man-made infrastructure, power supplies, and available workforce is also provided. Current information on other nearby operators lets users gauge competitive commitments, to work on strategies for bidding, and to access details on well development and lifecycle for production timescales.

Geological details of fields in South America.

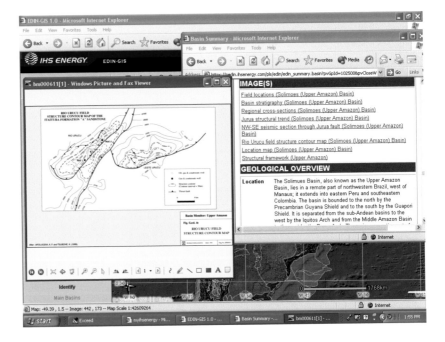

The EDIN-GIS service provides a browser-based interface through which data can be viewed and analyzed and which permits data to be served to desktop clients such as the ArcGIS family, so that it can be integrated with a client's own data sets. In addition, an export service is also provided (powered by Safe Software's FME), so the user can extract and export data selected through the EDIN-GIS interface in a variety of other third-party proprietary formats.

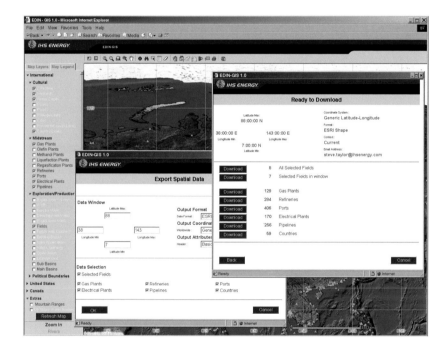

The EDIN-GIS Web Service permits data to be linked directly from clients' systems or downloaded in a variety of commonly used formats.

Behind the EDIN-GIS is a cluster of Sun UNIX machines that function as the Web, map, database, and application servers. The overall development framework is based on J2EE, with Apache Tomcat technologies used for Web and application server development. Spatial data is currently managed in an Oracle 8.1.7 database through ArcSDE 8.1. Novell iChain® and eDirectory™ software provide user authentication and access security for both local and remote uses. EDIN-GIS is built on ArcIMS with user interface development, and query tools being written in Java.

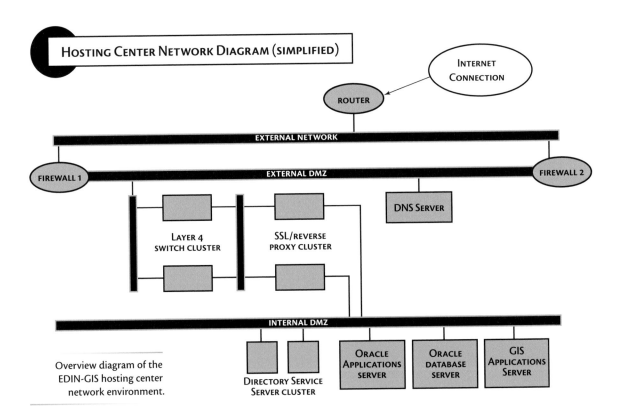

Overview diagram of the EDIN-GIS hosting center network environment.

FUTURE

The implementation of EDIN-GIS is, in fact, part of a wider Web architecture vision for IHS Energy. This includes enabling Web access to other databases maintained by the company, and developing a series of common services which, utilizing SOAP and XML standards, can be directly integrated with a client's own applications. Such services include generic query, graph, report, export, and map services functions. In the oil industry, where many companies have large investments in existing, often proprietary, systems they wish to maximize, such component services will allow direct access to the wealth of data and functionality in IHS Energy's servers. With minimal development, legacy systems can be enhanced by directly embedding remotely hosted services. The client will not only be freed from managing and maintaining massive data sets locally, but data and functionality can also be accessed from familiar systems that reduce the learning curve, the need for investment in new hardware and software, and system maintenance. For IHS Energy, as well, component Web services offer new ways to serve content, support clients, and open up new markets.

Home on the Web

12

ONE OF THE FIRST commercial Web sites to introduce a GIS service was REALTOR.com®, the official site of the National Association of Realtors (NAR) in the United States, which provides prospective homebuyers access to a nationwide database of over two million property listings. The original, innovative GIS capability in 1997 let users look for a potential new home using simple spatial criteria—to find only three-bedroom, three-bath, single-story homes in a particular area, for example—and then to generate dynamic location maps of the results. Maps were automatically centered on the selected property, and showed basic road network and topographic features. They could be zoomed in and out, and printed.

Since then, Homestore, Inc., which manages the REALTOR.com site and a number of related property sites, all accessible via the Web, has continued to expand the use of GIS Web services throughout its sites and now boasts some of the most sophisticated Web Services available. The evolution of Homestore's GIS services over the last six years illustrates the way GIS Web services generally have developed, from relatively simple mapping services hosted on individual machines, to the present day—where data and functional services can be pulled from multiple servers, be combined and integrated, and provide a degree of flexibility and functionality that was never possible before.

INFORMATION WANTED

People looking for a new home immediately become voracious consumers of information. It is easy to understand why. Buying a new house is often the biggest investment a family ever makes and so it is one they will make only after considerable amount of study. In addition, homebuyers are often moving to a new and unfamiliar city or town, and so are full of questions: *Is the neighborhood safe? How are the local schools? Will the commute to work be snarled with traffic?* In addition, they must contend with the pressure of uncertainties: new jobs to start, new schooling to arrange and, for those selling their old home, the timing of that sale in relation to the buying of the new one.

The argument for Web-based property listings is therefore compelling. A Web-based property listing such as one found on the Homestore site can provide a wealth of the kind of information a homebuyer wants and needs. It permits access to the listings from multiple agencies simultaneously, and provides easy search facilities so that properties can be selected based on their particular characteristics and price range. It can also lighten some of the workload, because options can be compared online, new neighborhoods explored, and a list of potential homes assembled, all before anyone steps out for the first reconnaissance. What is more, the ease with which different properties can be reviewed and compared means more choices can be considered. In short, Web services lower the barriers to knowledge considerably.

The growth of the REALTOR.com Web site testifies to this demand. When launched in 1996, the company reported holding about 500,000 property listings, with one million unique users visiting the site in 1997. By early 2003, there were more than two million property listings and five million unique visitors. Over the last five years Homestore has added a number of related sites that serve particular market segments. These include listings of newly constructed homes and developments, apartments and rentals, corporate housing, and sheltered or senior housing.

The map features added to the REALTOR.com Web site in 1997 were, at the time, state of the art. They were deployed using custom Internet Map Server (IMS) technology developed by ESRI (the forerunner of ESRI's current flagship Internet mapping technology, ArcIMS), and could generate dynamic maps quickly using very detailed national databases. Spatial data was managed in ArcSDE and stored in an Oracle database. ESRI developed a nationwide map service built on GDT Dynamap/2000®

Homestore is on the Web at www.homestore.com.

The Homestore site is a mine of information for those looking for a new house. The site includes more than two million property listings across the United States and Canada, and details of homes for rent, housing for the elderly, and other related services.

street data and other sources, and it enabled users to view detailed street maps of any neighborhood in the United States. The same street data was used by REALTOR.com to automatically geocode homes for sale, based on their ZIP Code or address.

The system permitted users to zoom on a particular area of interest and to orient themselves with maps showing neighborhood boundaries. Because all properties were geocoded, flexible spatial queries to identify homes for sale in a given area could be added to the list of search criteria. On finding a property that looked interesting, street maps of the local area could be generated, showing the location of the property in question, road network, street names, and related details.

The street map service proved extremely successful and in 1998 was augmented by a simple "find neighborhood" application. This enabled the user to specify the type of neighborhood they wanted to live in, either by grading some different characteristics or by identifying a neighborhood they liked by ZIP Code. A map of neighborhoods shaded to indicate how well they matched the specified criteria would be returned.

These mapping functions were robust, efficient, and worked well. However, as Web technology developed and demand both for the kinds of data served and the way it was presented became more sophisticated, an upgrade was soon required.

On the old Web site, users began with this national view and then drilled down to a neighborhood of interest.

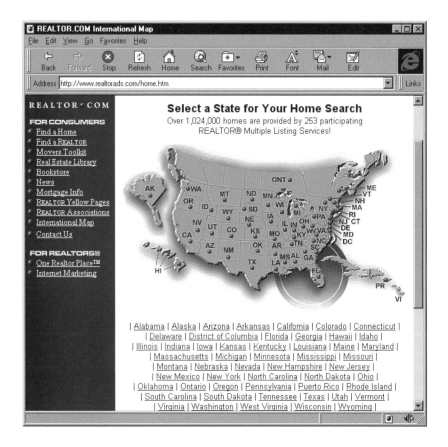

The popularity of the site's mapping tools highlighted their utility and relevance to many other parts of the system. Many of the questions homebuyers had revolved around location—proximity to schools and services, demographics, tax zones, and areas prone to flooding or earthquakes. All these factors could be mapped, and would make the site even more attractive to homebuyers if they could be easily accessed at the same time as listings were being viewed. New and improved spatial data sets were becoming widely available, as were sophisticated routing algorithms that could be used to help provide users with route plans and driving directions to get from one property to the next.

A NEED TO REMODEL

The original system was, however, poorly equipped to meet this rapidly evolving demand. The custom IMS application provided great flexibility and functionality for developers in 1997, but was beginning to be outpaced by new technology. Any enhancements required amendment and recompilation of the entire application. In addition, because it was developed in the days before XML became the industry standard for formatting Web communication between applications, all messages sent from browsers to the system had to be sent as coded HTML pages. This worked reliably, but added an extra layer of complexity to programming and integration with other applications. Adding new spatial layers such as detailed topographic mapping, census statistics, and neighborhood reports would require significant investment in upgrading hardware and time-consuming database tuning. While development and expansion was possible, the costs and potential disruption involved in carrying these through represented a significant overhead.

In addition, Homestore was expanding. As it added new sites focusing on different segments of the property market—such as rentals and retirement homes—there was a realization that each could benefit from mapping functionality similar to that provided in the REALTOR.com site. These sites added later, however, were being developed with the latest Web technology and there was little enthusiasm for adding spatial functionality if it entailed having to interface with the highly customized IMS application.

As a result, in late 2002 a decision was made to remodel spatial functionality on distributed Web Services architecture. This offered flexibility: the capability of tapping into the rapidly increasing number of professionally hosted geographic data sets, but without the overhead of in-house storage and management. In addition, a sophisticated range of spatial services that provided both existing and added functionality was available. These new services not only reduced the effort involved in system development, testing, and ongoing maintenance, but, based on XML and SOAP standards, they could be easily integrated into other components of the system, making rapid deployment of new, flexible mapping applications a real possibility.

ESRI's ArcWeb services was selected to provide all spatial data and functionality on Homestore sites. While the core property data and application servers remain in Homestore's offices in Westlake Village, California, mapping is provided through ESRI's hosted facilities over one hundred miles away in the city of Redlands. Not only are the map services hosted on machines dedicated solely for the provision of spatial data and functionality, they come with round-the-clock support and maintenance from GIS specialists, and are backed by the security of a full back-up system located in Mesa, Arizona.

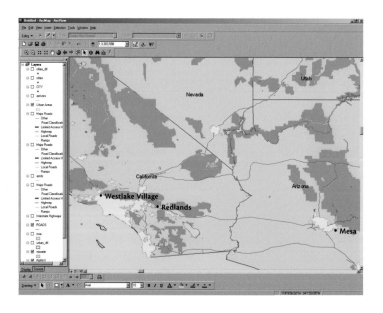

This system opens up a huge diversity of data sets drawn from a very wide range of sources. Rather than the relatively simple topographic and road map features that comprised mapping in the earlier system, available data sets now include shaded relief, land cover, population density and demographics, administrative boundaries, detailed transportation network, and information on hazards such as flooding. Data is hosted locally on ESRI servers or is provided by third-party hosting services accessed through the Geography Network architecture. This configuration removes not only the need to manage these data sets internally, but also means that maintenance is done at the source by the service provider, ensuring that it is always current.

Application services are made available using the XML-based SOAP protocol and can be easily integrated with any Web-enabled applications. The functionality includes the ability to locate places by city or town name, to geocode addresses based on ZIP Code or address details, to establish the shortest route between two points, to generate predefined and user-customized maps images, and to manage the upload and search of custom points-of-interest (POI) tables. Previously, adding new spatial functionality to the Homestore Web site was a major exercise, requiring new code to be written, tested, and carefully integrated with existing applications. New functionality can now be easily snapped into place with minimal disruption.

MORE THAN JUST A NEW COAT OF PAINT

The look and feel of the REALTOR.com and related sites have largely been preserved from its earlier incarnations, but the user will quickly become aware both that the map data sets and functions are richer than before, and that they are far more deeply integrated with the basic functionality of the site. Thus, for example, it is now possible to move from overview maps showing state and county boundaries to maps showing topography as shaded relief, river systems, and lakes and major roads. Larger-scale maps not only show the transportation network, but also local boundaries, commercial districts, parkland, and the location of schools. Properties can be mapped individually, or, if a number have been selected, they can be mapped together, giving a better idea of their location relative to each other. Driving directions, distances, durations, and route maps can also be generated from a given location to a selected property, allowing users to plan journey durations and routes more accurately.

Homes can be searched by map, ZIP Code, type of neighborhood, property features, or cost. Once a property is selected it can be mapped and road plans and driving directions generated.

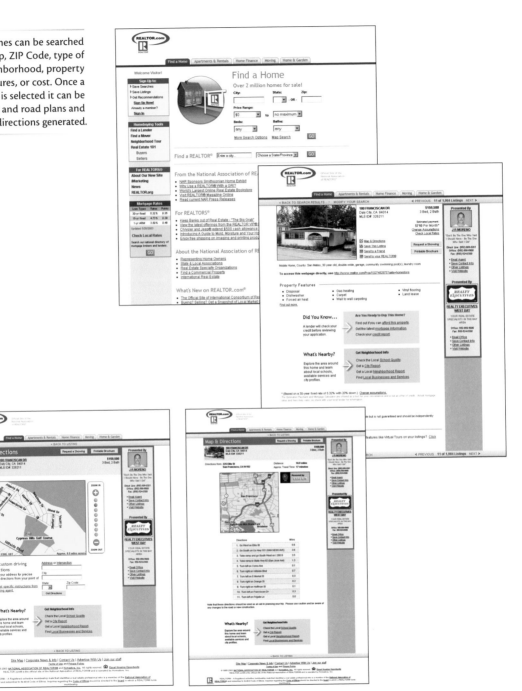

Spatial functionality is also increasing the range of information that can be provided about neighborhoods. Having found a property of interest, a user can now have the application select the nearest schools within the area and access detailed information about those schools from databases compiled by such companies as National School Reporting Services, Inc. The database can include such details as student-to-teacher ratios, facilities, and average scores on standardized tests such as the Scholastic Aptitude Test (SAT). Moving on to the neighborhood itself, details can be retrieved about the average size, age, and selling price of properties in the area, and the lifestyle profiles of the communities living there—including average age, salary, level of education, unemployment rate, and percentage of home ownership. Checking the local facilities is also easy. Details

These images show summary details of the neighborhood in which the property is located, and retail stores in the surrounding area. The Web service automatically calculates distances from the selected property to nearby stores, which can be used to map them.

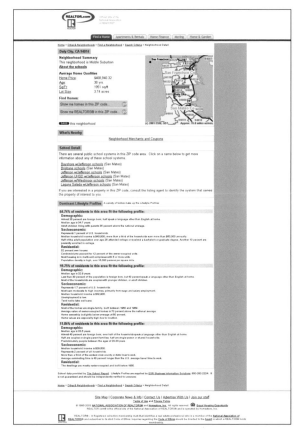

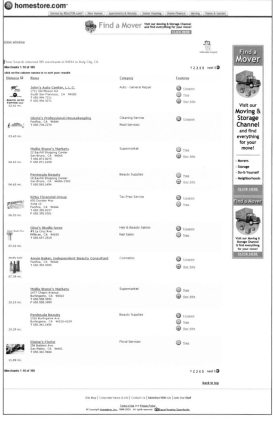

on local shops, hospitals, restaurants, and other services within the area can be retrieved at a click of a button. Each is listed with an address and distance from the selected property, in addition to contact details and a link to an Internet site, if one is available. Each can, of course, be mapped in relation to the selected property. Finally, to compare a selected property with others in the area it is possible to select and review all currently available property in the neighborhood. Similar data and functions are now being deployed across all of Homestore's sites.

Providing this kind of level of service back in 1997 would have been almost unthinkable, and, if it was contemplated, would have involved massive investment in application development, hardware, software, network infrastructure, and then ongoing maintenance and support. That it has been brought to fruition in 2002 and that it took just weeks to move from initial design to final release are testimony to the functionality and flexibility of modern Web services.

Homestore and the REALTOR.com site have followed the development of GIS Web services from their early days, as basic server applications communicating with Web browser clients through coded HTML messages, to a point where the Web not only connects servers with clients, but also, supported by internationally recognized standards, it connects services with services, greatly expanding the range and complexity of functionality that can be offered. In the end, this makes it easier to deliver targeted, appropriate services to the user as and when they are needed. While final decisions on a home will probably always be made after opening a front door, walking into an airy kitchen or sunlit living room, or strolling around the garden, Web services such as Homestore's mean that such decisions can be made with confidence—backed by detailed information and research done ahead of time.

Afterword

In discussions with those involved in developing and managing the GIS Web services described in this book, one thing is clear—this is only the beginning. For many users, the services currently offered are only the start of much larger plans to bring Web services technology to many more areas of their operations.

The business case for Web services is persuasive:

▸ Data and services can be released to a worldwide audience, permitting users to access them when and where they are needed;

▸ Duplication of both data and functionality among sites can be reduced;

▸ Overhead for data storage, maintenance, and for development of functionality is reduced;

▸ Services are compatible and can be easily integrated with each other and other technologies;

▸ Service providers reach multiple clients and achieve economies of scale and specialization that can be invested to improve service quality and reliability;

▸ Users can select from multiple service providers, improving competition and cost-effectiveness.

The case makes as much sense for systems within individual organizations as it does for those that span the world—facilitating data and system management and improving how data, applications, and functions are brought to the user and integrated with the tools of everyday life. It is particularly relevant in relation to geographic information and analysis, which has, in the past, been difficult to distribute and which requires specialized skills and resources to analyze and manage effectively.

That said, it is also clear that the technology behind Web services is still maturing. The key standard on which they are based, XML, continues to be debated, reviewed, and developed. The latest recommended version, XML 1.1, was submitted to W3C in October 2002 and work continues on enhancements—for example to develop an XML Query Language. The same is true for other standard protocols—SOAP 1.2 has been under consideration by W3C since June 2001, and work is underway on enhancements and modification of WSDL, UDDI, and OGC's GML. Though XML and SOAP have achieved widespread acceptance, some initiatives such as UDDI have not, and may be superceded by other technologies. Despite increased use and performance of HTTPS networks, there continue to be concerns over security, performance, and manageability of services, which some of the enhancements are designed to address.

What is clear, however, is that there has been a fundamental change in the way the Web is used. In the past, data that had been presented on a Web page could only be discovered and viewed; it was difficult to integrate the data with information from other Web pages or from locally hosted systems, and it was impossible to run applications across the Web. GIS Web services permit users to search, query, and retrieve Web-hosted databases and to integrate the results with other data, whether locally or remotely held. They enable users to access and use Web-hosted spatial functionality—to run proximity searches, geocoding, spatial modeling, or thematic reporting across the Internet. The Web is no longer a tool for passive presentation and retrieval of information; it has become a framework through which data can flow, be combined, selected, queried, and analyzed to create new information—new ways of presenting or analyzing data, new ways of looking at the world, new solutions, new insights.

The implications of this new framework for a number of organizations have been illustrated in this book. But what are the wider implications for the GIS industry as a whole?

The Web services framework provides the potential for centrally managed data sets and functionality to be accessed by a far wider audience and in far more sophisticated ways than has been possible in the past. As a result, greater attention will need to be focused on developing central server-side technology to support this. A new breed of GIS servers designed specifically for LAN and Internet clients can be expected. These enterprise servers will be able to support data management and complex analytical or editing tasks that can be undertaken equally within "thick" traditional desktop clients or through light Web-services clients. Further work will be required on Web protocols such as ArcXML, SOAP, and GML to enhance them, and to make them capable of delivering a level of presentation and sophistication that has not been seen to date—truly high-quality cartography, the ability to pass data and parameters to complex three-dimensional models and visualization tools, the ability to bring a number of models hosted at different remote locations together, to work on data that may be collected at other remote sites.

The changes will not only impact GIS technology. The ability to use externally hosted services for both common data and functionality will bring GIS to new audiences that have hitherto been put off either by lack of data or expertise, or by the overheads involved in establishing and managing systems locally.

Existing users will also benefit from access to externally hosted services that will greatly reduce data and system management overheads. Users can concentrate on the relatively few unique services or data sets within their particular organizations rather than having to worry about maintaining basic, generic mapping and data services. Data suppliers, whether private or government organizations, will need to focus far more closely on their ability to host robust, reliable Web services, but in return will find new solutions to issues such as distribution and management of updates and licensing, and copyright control.

To reiterate a point from chapter one, GIS Web services rely entirely on accepted, open standards. Open standards can be time-consuming to work out and require trust and shared effort and willingness on the part of many different, sometimes competing, organizations. But as we witness everyday, the results can be dramatic. Take, for example, telecommunications, and the complexity involved in picking up a mobile phone in the United States and connecting within seconds to the mobile phone of a friend or business associate in South Africa: your conversation passes

through a network composed of many different components provided by many different vendors—what makes it all work are open, accepted standards. As the standards that support GIS Web services develop and are embraced by the GIS community, they will begin to form a global network that permits geographic data and functionality to flow around the world with similar ease.

▸ Book design, production, and image correction by Jennifer Galloway
Book cover design by Doug Huibregtse
Copyediting by Tiffany Wilkerson
Printing coordination by Cliff Crabbe